Francis Edmund Anstie

Neuralgia

And the Diseases that Resemble it

Francis Edmund Anstie

Neuralgia
And the Diseases that Resemble it

ISBN/EAN: 9783337035266

Printed in Europe, USA, Canada, Australia, Japan

Cover: Foto ©berggeist007 / pixelio.de

More available books at **www.hansebooks.com**

AND

THE DISEASES THAT RESEMBLE IT.

BY

FRANCIS E. ANSTIE, M.D., LONDON,

FELLOW OF THE ROYAL COLLEGE OF PHYSICIANS; HONORARY FELLOW OF KING'S
COLLEGE, LONDON; SENIOR ASSISTANT PHYSICIAN TO WESTMINSTER
HOSPITAL; LECTURER ON MEDICINE IN WESTMINSTER
HOSPITAL SCHOOL; PHYSICIAN TO THE BEL-
GRAVE HOSPITAL FOR CHILDREN.

NEW YORK:
BERMINGHAM & CO., UNION SQUARE.
1882.

W. L. MERSHON & CO.,
Printers, Electrotypers and Binders,
RAHWAY, N. J.

PREFACE.

I BELIEVE it will not be disputed that there was considerable need for an English treatise dealing rather fully with the subject of Neuralgia, and therefore I hope that the profession will be willing to give me a hearing. The present work, moreover, does not profess to be a mere compilation of standard authorities corrected down to the present time, but puts forward a substantially new view of the subject—at least, a view that has been only briefly sketched by me in an article that appeared, three years ago, in Reynolds's "System of Medicine." My principal object, in writing this volume, was to vindicate for Neuralgia that distinct and independent position which I have long been convinced it really holds, and to prove that it is not a mere offshoot of the Gouty or Rheumatic diatheses, still less a mere chance symptom of a score of different and incongruous diseases. In order to set the diagnosis of true Neuralgia from its counterfeits in the clearest light, it seemed advisable to draw separate pictures of each of the latter (at least of as many as are of real importance) and present them separately, as a kind of gallery of spurious neuralgias, and this I have done in the second part of the volume. No one who had not tried to do it would imagine how difficult this latter kind of work is. It was necessary for the sketches to be very brief (unless my book was to become unmanageably large), and yet to be as truthfully characteristic as possible; and it was necessary also that only those diseases which so much resemble Neuralgia as practically to lead medical men astray in diagnosis, should be dealt with. The selection of the subjects, and the execution of this part, took a long time, though it only covers about fifty pages. Then, as regards Neuralgia itself, it became necessary to completely recast the chapters on "Pathology" and on "Complications," on account of some of the polite criticisms which Dr. Eulenburg directed (in his recent "Lehrbuch der Nervenkrankheiten ") to my argument in the article above referred to, since it was obvious that a too brief statement of my views had caused them to be partially misunderstood by the German physician. These chapters (Part I., Chapters II. and III.) are certainly the most important portion of my book, and I would particularly direct attention to them, in order that their contents may be affirmed or corrected: the reader will at any time find that they contain a kind of investigation never before systematically carried out with regard to Neuralgia. The causes above mentioned, together with others over which I had no control, have kept back the appearance of this work so long beyond the date for which it was originally announced, that I feel I ought to apologize for an amount of delay that would seem hardly justified by the moderate size of the volume.

16 WIMPOLE STREET, LONDON, *October* 1, 1871.

CONTENTS.

INTRODUCTION—ON PAIN IN GENERAL . . 7

PART I.
ON NEURALGIA

	PAGE
CHAP. I.—CLINICAL HISTORY	12
II.—COMPLICATIONS OF NEURALGIA . . .	79
III.—PATHOLOGY AND ETIOLOGY OF NEURALGIA.	96
IV.—DIAGNOSIS AND PROGRESS OF NEURALGIA .	142
V.—TREATMENT OF NEURALGIA	149

PART II.
DISEASES THAT RESEMBLE NEURALGIA.

CHAP. I.—MYALGIA	196
II.—SPINAL IRRITATION	200
III.—THE PAINS OF HYPOCHONDRIASIS . .	207
IV.—THE PAINS OF LOCOMOTOR ATAXY . .	210
V.—THE PAINS OF CEREBRAL ABSCESS . .	213
VI.—THE PAINS OF ALCOHOLISM. . . .	215
VII.—THE PAINS OF SYPHILIS	218
VIII.—THE PAINS OF SUBACUTE AND CHRONIC RHEUMATISM	225
IX.—THE PAINS OF LATENT GOUT . . .	227
X.—COLIC, AND OTHER PAINS OF PERIPHERAL IRRITATION	229
XI.—DYSPEPTIC HEADACHE	231

INTRODUCTION.

ON PAIN IN GENERAL.

Although it is, in a general way, unadvisable to introduce abstract discussions into a treatise which should be strictly practical, it is almost impossible to avoid some few general reflections on the physiological import of Pain, as a preliminary to the discussion of the maladies which form the subject of this volume. This whole group of disorders is linked together by the fact that pain is their most prominent feature; and, with regard to most of them, the relief of the pain is the one thing required of the physician. It seems, therefore, very important that we should ascertain, at least approximately, in what the immediate state consists, which consciousness interprets as pain. It is not necessary to enter at this stage into any inquiry as to the pathological causes of the phenomenon; what we know of these, and it is unfortunately too little, will be discussed in detail under the headings of the several affections which I shall have to describe.

The question before us now is this : What is that functional state of the nerves which consciousness interprets as pain ? Is it, or is it not, an exaltation of the ordinary function of sensation ?

The latter question is generally answered affirmatively, without much thought, by those to whom it casually occurs; but indeed there is plenty of prescriptive authority for so dealing with it. Pain has been described by some of the most distinguished writers on nervous diseases as a hyperæsthesia. Yet there is really little difficulty in convincing ourselves, if we institute a thorough inquiry into the matter, that pain is certainly not a hyperæsthesia, or excess of ordinary sensory function, but something which, if not the exact opposite of this, is very nearly so.

The leading fallacy in the common view is the confusion which is perpetually being made between function and action. Now, the function of individual nerves is very nearly a constant quantity, at least, it varies only within narrow limits ; while the action of the same nerves may be almost any thing. The function of the nerve is that kind of work for which it is fit when its molecular structure is healthy; it is the series of dynamic reactions which are necessarily produced in nerve-tissue by the external influences which surround and impinge upon it in the conditions of ordinary existence. The action of nerves, under the pressure of extraordinary influences, may include all manner of vagaries which really have nothing in common with the effects of ordinary functional stimulation; which are, in fact, nothing but perturbation. No one can suppose, for instance, that the explosive disturbances of nerve-force which give rise to the convulsions of tetanus are any mere exaggerated degree of the orderly and symmetrical action by which the healthy nerve responds to the stimulus of volition ordering a given set of muscles to contract; they are something quite different in kind. And so it is with the sensory nerves. The functions of these conductors, in health, is to convey to the perceptive centres the sensations, varying only within a most limited range, which correspond to a state of well-being of the organs, and which excite only those reflex actions that are necessary to life. Thus the large surface of sensitive nerve terminals which is represented by the collective peripheral branches of the fifth cranial conveys to the medulla oblongata an impression, derived from the temperature and movement of the surrounding air, when the latter is neither too hot nor too cold, which imparts to the brain a perception of comfortable sensations, and excites in return the reflex action of breathing, which is necessary to life. But the impression produced on this same peripheral expanse of nerve-branches by prolonged exposure to cold wind may, and often does, convey to the centres sensations which are quite different and provokes reflex movements which are altogether abnormal. Pain is the product in one direction ; sneezing, perhaps, in the other. It seems absurd to say that sneezing is any part of the function of those motor nerves whose action regulates the performance of expiration. And it appears to me not less absurd to say that pain is the function of the sensitive fibres of the trigeminus. But the best way, perhaps, to illustrate the looseness and incorrectness of applying the term "hyperæsthesia" (implying exalted function) to the state of sensitive nerves when suffering pain, is to examine the condition of distinctive perception in the very same parts to which the painful nerves are distributed. It will invariably be found, as we shall have occasion to see more fully proved hereafter, that, in parts which are acutely painful, a marked bluntness of the tactile perceptions can be detected. The tactile perceptions are, no doubt, conveyed by an independent set of fibres from those which convey the sense of pain.* Yet it is surely impossible to believe the effect of the same influence, in functional power can be different—much more than it can be exactly opposite—in the two cases.

---

* See, on this subject, some remarks, in my work on "Stimulants

If pain be not a heightening of ordinary sensation, then we seem to be shut up to the idea that it is a perversion owing to a molecular change of some part of the machinery of sensation which frustrates function. For it is to be observed that, while the sensations conveyed by the healthy nerve are correct in the indications which they afford to the percipient brain, the indications given by pain are vague and untrustworthy, and often seriously misleading. Not to speak of the nerves of special sense, or of the fibres which convey the sensations of muscular movement, even the nerves of common sensation do carry to the internal perception, in health, a distinct impression of the well-being of the organs to which they are distributed. Mr. Bain † has well pointed out the positive character of this feeling, which is so often incorrectly referred to as if it were a mere negation of feeling. It is a sensation of equable and diffused comfort, if I may be allowed to use the expression, which streams in from all parts of the organism; and there is no possibility of comparing it, in any scale of less or more, with the sensation of pain; for the latter commonly conveys no correct information as to the organ from which it proceeds, or appears to proceed. Especially is this the case in the neuralgias, for more commonly than not the apparent seat of the pain is widely removed from the actual seat of the mischief which causes it.

If we inquire a little further into the circumstances under which various kinds of pain occur, we gain some fresh suggestions. Among the neuralgias, those are the most acutely agonizing which occur under circumstances of impaired nutrition incident to the period of bodily decay, and strong reasons will be hereafter adduced for the belief that there is especial impairment of the nutrition of the central end of the painful nerves. To find a parallel to the severity of this kind of pains we must turn to the case of organic tumors, which, from their position, structure, and mode of growth, necessarily exercise continuous and severe pressure on the branches or the trunk of a nerve; or to the class of pains which attend severe cramp, or tonic contraction of muscles. Now, it can scarcely be doubted that in the latter instance there is an abnormally rapid and violent destruction of tissue going on; at the very least there is an extraordinarily violent and irregular manifestation of motor force. In any case the patent fact here is dynamic

---

and Narcotics" on Sir W. Hamilton's "Theory of the Relations of Perception and Common Sensation."

A very distinct and careful statement of the distinction between pain and hyperæsthesia will be found in a prize essay "On Neuralgia" by M. C. Vanlair, Jour. de Bruxelles, tom. xl., xli., 1865.

† "Senses and Intellect."

perturbation of a severe kind; and, in the instance of organic tumors exercising steady and continuously increasing pressure on nerves, one can scarcely doubt that a similar perturbation, less intense but more enduring, is necessarily set up. That which can be done in the way of producing severe pain by these severe affections of the peripheral portions of nerves, or of tissues lying outside them, we might *a priori* expect would be effected by slighter but continuous changes in the nutrition of the more important portion of the nerve itself—its central gray nucleus. One would say that a pathological process which continuously and progressively lowered the standard of nutrition here must interfere from hour to hour, certainly from day to day, with that regular and equable distribution of force which is the essence of unimpeded function.

Take, again, the case of the very severe pain which frequently attends inflammation of the pleura and of the peritoneum. Whatever theory of the causation of these pains we may adopt, it is certain that one most important element in their production and maintenance is the continual movement and friction of the affected parts. But there is little doubt that the moving muscles are involved in the inflammatory process, as Dr. Inman has correctly observed. It would seem plain that under these circumstances—an inflamed muscular structure forced to perform its ordinary contractions as well as it can—there must be powerful dynamic perturbation going on.

If perturbation of nerve-function,—a disturbance quite different from mere exaltation of the normal development of nerve-force—be the essence of pain, how comes it that pains of the severest type may be produced by changes in structures which are usually described, for practical purposes, as lying outside the nervous system ? We must, in the first place, remark that the externality of any bodily tissue to the nervous system is more apparent than real. Microscopic researches are constantly revealing nerve-fibres, in ever-increasing profusion, which penetrate to parts seemingly the least vitalized in the organism. But, in any case, the nerves are certainly the ultimate channel of communication between the suffering part and the sentient centre. It seems, therefore, the inevitable conclusion that a dynamic perturbation going on in the non-nervous tissue is continued along the nerves themselves : and that the severity of the pain perceived by the conscious centres is proportionate to the tumultuousness, the want of coordination, and the waste with which force is being evolved in the cramped muscle, or whatever structure it may be, in which the pain takes its source.

Not to pursue these topics further, we may sum up the considerations which have now been adduced, in the following general propositions, which will tend to simplify the examination of the various painful disorders which we are about to discuss :

1. Pain is not a true hyperæsthesia; on the contrary, it involves a lowering of true function.
2. Pain is due to a perturbation of nerve-force, originating in dynamic disturbance either within or without the nervous system.
3. The susceptibility to this perturbation is great in proportion to the physical imperfection of the nervous tissue, until this imperfection reaches to the extent of cutting off nervous communications (paralysis).

PART I.

ON NEURALGIA

CHAPTER I.

CLINICAL HISTORY.

Neuralgia may be defined as a disease of the nervous system, manifesting itself by pains which, in the great majority of cases, are unilateral, and which appear to follow accurately the course of particular nerves, and ramify, sometimes into a few, sometimes into all, the terminal branches of those nerves. These pains are usually sudden in their onset, and of a darting, stabbing, boring, or burning character; they are at first unattended with any local change, or any general febrile excitement. They are always markedly intermittent, at any rate at first; the intermissions are sometimes regular, and sometimes irregular; the attacks commonly go on increasing in severity on each successive occasion. The intermissions are distinguished by complete, or almost complete, freedom from suffering, and in recent cases the patient appears to be quite well at these times; except that, for some short time after the attack, the parts through which the painful nerves ramify remain sore, and tender to the touch. In old-standing cases, however, persistent tenderness, and other signs of local mischief, are apt to be developed in the tissues around the peripheral twigs. Severe neuralgias are usually complicated with secondary affections of other nerves which are intimately connected with those that are the original seat of pain; and in this way congestions of blood vessels, hypersecretion or arrested secretion from glands, inflammation and ulceration of tissues, etc., are sometimes brought about.

The above is a general description of neuralgia which will identify the disease sufficiently for the purpose of introducing it the attention of the reader. We must now proceed to give a more accurate account of its

Clinical History and Symptoms.—These vary so greatly in different kinds of neuralgia that it will be necessary to discuss the greater part of this subject under the headings of the special varieties of the disease. There are certain common features, however, in all true neuralgias.

I. In the first place, it is universally the case that the condition of the patient, at the time of the first attack, is one of debility, either general or special. I make this assertion with confidence, notwithstanding that Valleix, and some other very able observers, have made a contrary statement. In the first place, it is certainly the case that the larger half of the total number of cases of neuralgia which come under my care are either decidedly anæmic, or else have recently undergone some exhausting illness or fatigue; and if other writers have failed to see so many neuralgic patients in whom these conditions were present, it must certainly be because they have limited the application of the term "neuralgia" within bounds which are too narrow to be justified by any logical argument; as will, indeed, be shown at a later stage. On the other hand, although a considerab e number of neuralgic patients have an externally healthy appearance, as indicated by a ruddy complexion and a fair amount of muscular development, it cannot be admitted that these appearances exclude the possibility of debility, either structural or functional, of the nervous system. The commonest experience might teach us that, so far from the nervous system being invariably developed with a corresponding completeness and maintained with a corresponding vigor to those which distinguish the muscular system and the organs of vegetative life, there is often a very striking contrast between these in the same individual. What physician is there who has not seen epileptic patients, in whom mental habitude, a low cranial development, imperfect cutaneous sensibility, and other obvious marks of deficient innervation, were marked and striking features at, or even before, the first occurrence of convulsive symptoms, while the body was robust, the face well colored, and the muscular power up to or beyond the average? Now, it will invariably be found, on carefully sifting the history of apparently robust neuralgic patients, that they, too, have given previous indications of weakness of the nervous system: thus, women, who, after a severe confinement attended with great loss of blood, are attached with *clavus hystericus* or with *migraine*; will inform us that whenever, in earlier life, they suffered from headache, the pain was on the same side as that now affected, and chiefly or altogether confined to the site of the present neuralgia. In a considerable number of cases, also, in which I have been able to observe accurately the events which preceded an attack of neuralgia, it has been found that the skin supplied by the nerves about to become painful was anæsthesic to a remarkable degree; and it is very often the case that a more moderate amount of blunted sensation was perceptible in these parts during the intervals between attacks of pain. A somewhat delusive appearance of general nervous vigor is often conveyed to the observer of neuralgic patients, by reason of the intellectual and emotional characteristics of the latter. Both ideation and emotion are, indeed, very often quick and active in the victims of neuralgia, who in this respect differ strikingly from the majority of epileptics. But this mobility of the higher centres of the nervous system is itself no sign of general nervous strength ; which last can never be possessed except by those in whom a certain balance of the various nervous functions is maintained. Much more will be said on this topic when we come to discuss the etiology of neuralgia. Meantime I may content myself with repeating the fact which is indubitably taught by careful observation—that neuralgics are invariably marked by some original weakness of the nervous system; though in some cases this defect is confined strictly to that part of the sensory system which ultimately becomes the seat of neuralgic pain.

Another circumstance is common to all neuralgias of superficial nerves; and, as a large majority of all neuralgias are superficial in situation, this is, for practical purposes, a general characteristic of the disease. I refer to the gradual formation of tender spots at various points where the affected nerves pass from a deeper to a more superficial level, and particularly where they emerge from bony canals, or pierce fibrous fasciæ. So general is this characteristic of inveterate neuralgias, that Valleix founded his diagnosis of the genuine neuralgias on the presence of these painful points. Herein he appears to me to be decidedly in error. I have watched a great many cases (of all sorts of varieties as to the situation of the pain), and I have uniformly observed that in the early stages firm pressure may be made on the painful nerve without any aggravation of the pain; indeed, very often with the effect of assuaging it. The formation of tender spots is a subsequent affair: they develop in those situations which have been the foci, or severest points, of the neuralgic pain. There is however, a point which, though not always, nor often, the seat of spontaneous pain, is nevertheless very generally tender. Trousseau, who criticises unfavorably the statement of Valleix as to the situation of the points douleureux, insists that this tender spot, which is over the spinous processes of the vertebræ corresponding to the origin of the painful nerve, and which he calls the points apophysaire, is more universally present than any of those pointed out by Valleix. I shall hereafter endeavor to show that these spinal points are by no means characteristic of neuralgia; they are present in a variety of affections which were ably described, under the heading of "Spinal Irritation," many years ago, by the brothers Griffin. ["Observations on the Functional Affections of the Spinal Cord," by William and Daniel Griffin. London, 1834] and they are also present with misleading frequency in cases of mere myalgia, such as I shall have to describe at a later stage.

Another characteristic of neuralgic patients in general is, I believe, a certain mobility of the vaso-motor nervous system and of the cardiac motor nerves; but I insist less on this than on the above-named features, because a more extended experience is necessary to establish the fact with certainty. Within my own experience it has always seemed to be the case that persons who are liable to neuralgia are specially prone to sudden changes of vascular tension, under emotional and other influences which operate strongly on the nervous system. The observation of this fact has been made accidentally, without any previous bias on my part, in the course of a large number of experiments made upon individuals free from manifest disease at the time, with Marey's sphygmograph.

Neuralgic attacks are always intermittent, or at the least remittent, in every stage of the disease.

The manner in which neuralgic pain commences is characteristic and important. There is always a degree of suddenness in its outset. When produced by a violent shock, it may, and often does, spring into full development and severity at once, of which, perhaps, the most striking example is the sudden and violent neuralgic pain of the eyebrow which some persons experience from swallowing a lump of undissolved ice. Usually, however, the first warning is a sudden, not very severe, and altogether transient dart of pain. The patient has probably been suffering from some degree of general fatigue and malaise, and the skin of the affected part has been somewhat numb, when a sudden slight stitch of pain darts into the nerve at some point which corresponds to one of the foci hereafter to be particularized. It ceases immediately, but in a few seconds or minutes returns; and these darts of pain recur more and more frequently, till at last they blend themselves together in such a manner that the patient suffers continuous and violent pain for a minute or so, then experiences a short intermission, and then the pain returns again, and so on. These intermittent spasms of pain go on recurring for one or several hours; then the intermissions become longer, the pain slighter, and at last the attack wears itself out. Such is generally the history of first attacks, especially in subjects who are not past the middle age, nor particularly debilitated from any special cause.

A point of interest in connection with the natural history of the neuralgic access is the condition of the circulation. The commencement of pain is generally preceded by paleness of skin and sensations of chilliness. At the commencement of the painful paroxysm, sphygmographic observation shows that the arterial tension is much increased, owing, in all probability, to spasm of the small vessels. This condition is gradually replaced by an opposite state, the pulse becoming large, soft, and bounding, though very unresisting, and giving a sphygmographic trace which exhibits marked dicrotism. Simultaneously with this the skin becomes warmer, sometimes even uncomfortably warm, and there is frequently considerable flushing of the face.

The final characteristic common to all neuralgias is that fatigue, and every other depressing influence, directly predispose to an attack, and aggravate it when already existing.

Varieties.—It is possible to classify neuralgias upon either of two systems: first (*a*), according to the constitutional state of the patient; and, secondly (*b*), according to the situation of the affected nerves. It will be necessary to follow both these lines of classification, avoiding all needless repetition.

(*a*) In considering the influence of constitutional states upon the typical development of neuralgia, it will be convenient to commence with the group of cases in which the general condition of the organism produces the least effect. This is the case when the pain is the result of direct injury to a nerve-trunk, whether by external violence, by the mechanical pressure of a tumor, or by the involvement of a nerve in inflammatory or ulcerative processes originating in a neighboring part. As regards the development of symptoms, the important matters are, that the pain in these cases commences comparatively gradually, that the intermissions are usually more or less complete, and that the pain is far less amenable to relief from remedies, than in other forms of neuralgia. The little that can be said about the form which is dependent upon progressively increasing pressure, or involvement of a nerve in malignant ulcerations, caries of bones or teeth, etc., falls under the heads of Diagnosis and Treatment, and need not detain us here. The clinical history of neuralgia from external violence, however, requires separate discussion:

1. Neuralgia from external shock may be produced by a physical cause (as by a fall, a railway collision, etc.), which gives a jar to the central nervous system; or by severe mental emotion, operating upon the same part of the organism. Under either of these circumstances the development of the affection may occur at once, but by far the most frequently it ensues after a variable interval, during which the patient shows signs of general depression, with loss of appetite and strength. Sometimes vomiting, and in other instances paralysis, of a partial and temporary kind, occur. When once developed, the neuralgic attacks do not differ from those which proceed from causes internal to the organism. In the greater number of instances, so far as my experience goes, it is the fifth cranial nerve which becomes neuralgic from the effects of central shock. Illustrative cases will be given in the section on Local Classification. Meantime the important facts to note, in relation to the influence of constitutional states, are these: In the first place, the tendency of such accidents to excite neuralgia varies directly with the hereditary predisposition

evinced by the liability of the sufferer's family to neuralgic affections and to the more serious neuroses. Secondly, the likelihood of a neuralgic attack is indefinitely increased if he has already had neuralgia. Thirdly, although debility from temporary and special causes can rarely be sufficient to insure a true neuralgic access after a severe shock, it probably heightens, indefinitely, the tendency in a person otherwise predisposed. Delicate women are many times more liable to experience such consequences, from a physical or mental shock, than men of tolerably robust constitution.

2. Neuralgia from direct violence to superficial nerves is produced by cutting or, more rarely, by bruising wounds. Cutting wounds may divide a nerve-trunk (*a*) partially, or (*b*) completely.

(*a*) When a nerve-trunk is partially cut through, neuralgic pain occurs, if at all, immediately, or almost immediately, on the receipt of the injury. One such instance only has come under my own care, but many others are recorded. In my case the ulnar nerve was partly cut through, with a tolerably sharp bread-knife, not far above the wrist; partial anæsthesia of the little and ring fingers was induced, but at the same time violent neuralgic pains in the little finger came on, in fits recurring several times a day, and lasting about half a minute. Treatment was of little apparent effect in promoting a cure; though opiates and the local use of chloroform afforded temporary relief. The attacks recurred for more than a month, long after the original wound had healed soundly; and, for a long time after this, pressure on the cicatrix would reproduce the attacks. A slight amount of anæsthesia still remained, when I saw the patient more than a year after the injury.

(*b*) Complete severance of a nerve-trunk is a sufficiently common accident, far more common then is neuralgia produced by such a cause; indeed, so marked is this disproportion between the injury and the special result, that I have been led to infer that a necessary factor in the chain of morbid events must be the existence of some antecedent peculiarity in the central origin of the injured nerve. This opinion is rendered the more probable because the consecutive neuralgia is in some cases situated, not in the injured nerve itself, but in some other nerve with which it has central connections. Two such cases are recorded in my Lettsomian Lectures, [*Lancet*, 1866], in which the ulnar nerve, and one in which the cervico-occipital, were completely divided; in all three the resulting neuralgia was developed in the branches of the fifth cranial. Here we may suppose that the weak point existed in the central nucleus of the fifth; and that the irritation, or rather depression, communicated to the whole spinal centres by the wound of a distant nerve, first found, on reaching this weak point, the necessary conditions for the development of the neuralgic form of pain, which therefore would be represented to the mental perception as present in the peripheral branches of the fifth nerve. In all the cases which have come under my notice, the neuralgia set in at a particular period, namely, after complete cicatrization of the wound, and while the functions of the branches on the peripheral side of the wound were partly, but not completely, restored. The same obstinacy and rebelliousness to treatment are observed as in other instances of neuralgia from injury.

One of the cases above referred to may here be briefly detailed, as it shows very completely the clinical history of such affections. C. B., aged twenty-four, an agricultural laborer. applied for relief in the out-patient room of Westminster Hospital, suffering from severe neuralgic pains of the forehead and face of the left side. Then pains were felt in the course of the supra-orbital, ocular, nasal, and supra-trochlear branches, and also in the cheek, appearing, there, to radiate from the infra-orbital foramen. They had commenced about three weeks previously to the patient's first visit to the hospital, and about six weeks after the accident which appeared to have started the whole train of symptoms. This was a cutting wound, evidently of considerable depth as well as external size, toward the back of the neck, and so situated that it must have divided the great occipital nerve of the left side: and, from the man's account of the numbness of the parts supplied by the nerve which immediately followed the wound, there could be no doubt that this had occurred. There was no acute nerve-pain, either during the healing of the wound, which was rapid, or subsequently, until more than three weeks from the date of the injury; at this time there was still a considerable sense of numbness in the skin of the occipital and upper cervical region; but there now commenced a series of short paroxysms of pain in the forehead of the same side. These at first occurred only about twice daily, at regular intervals; the pain was not very sharp, and only lasted a minute or two. The attacks rapidly increased in frequency and duration, however, and extended their area. At the time when I first saw the case the pain was very formidable, it recurred with great frequency during the day, but would sometimes leave the patient free for several hours together. The site of the wound was occupied by a firm cicatrix of about a line in breadth and an inch and a quarter in length; pressure on this excited only a vague and slightly painful tingling in the part itself, but severely aggravated the trigeminal pains, or reproduced them if they happened to be absent. The regions supplied by the great occipital nerve were still very imperfectly sensitive. This patient gave me a great deal of trouble. He continued for many weeks under my care, and I can scarcely flatter myself that any of the numerous remedies which I administered internally, or applied locally, had any serious effect in checking the disorder. The subcutaneous injection of morphia gave some relief, as it always does, but this seemed to be perfectly transitory; and, although when the patient ceased to attend the hospital he was decidedly better, I cannot imagine that there was anything in it except the slow wearing out of the neuralgic tendency, very much without reference to the administration of any remedies.

The description of neuralgia from injury would be incomplete without some special words on a variety of this affection which has only very recently been described with that fulness which it deserves. I refer to the pains which are produced by gunshot injuries of nerves, received in battle, of which no sufficient account had been given until the publication of the experience of Messrs. Mitchell, Moorehouse, and Keen, in the late American civil war.*

From the interesting treatise of the above-named writers it appears that not merely is neuralgia of an ordinary type a frequent after-consequence of wounds, but that certain special pains are not unfrequently produced. In the more ordinary instances, pain is of the darting, or of the aching kind; and all writers on military surgery, who have recorded their experience of the results of wounds received in battle, have spoken of affections of this kind, for the most part singularly severe and obstinate, and in not a few recorded instances clinging to the patient during the remainder of his life. These pains may at times leave the sufferer, but they infallibly recur when from any cause his health is depressed, and it is an especially common thing for them to be evoked in full severity under the influence of exposure to cold, and particularly to damp cold.

But the American writers introduce us to another and more terrible neuralgia which is a, fortunately, less frequent result of serious injuries to nerves. They speak of it as a burning pain of intense and often intolerable severity; they believe that it seldom if ever originates at the moment of the injury, but rather at some time during the healing process; and it is especially noteworthy that it is sometimes felt not in the nerve actually wounded, but in some other nerve with which it has connections. After it has lasted a certain time, an exquisite tenderness of the skin is developed, and a peculiar physical change of skin-tissure occurs; it becomes thin, smooth, and glossy. It is a remarkable fact that these burning pains which are so definitely linked with a nutrition-change of skin are never felt in the trunk, and rarely in the arm or thigh, not often in the forearm or leg, but commonly in the foot or hand; and the nutrition changes of the skin are generally observed on the palm of the hand, the palmar surface of the fingers, or

---

* "Gunshot Wounds and other Injuries to Nerves." Philadelphia: Lippincott & Co., 1864.

the dorsum of the foot; rarely on the sole of the foot or the back of the hand. It is very interesting to remark that these skin-lesions correspond very nearly, not only to those observed in the cases of nerve-injury reported by Mr. Paget,[*] in which actual neuralgia was present (though the kind of pain is not exactly specified), but also very nearly with the nutritive changes observed by Mr. Jonathan Hutchinson in a number of cases of surgical injuries of nerves.[†] The tendency of neuralgic pain accompanied by nutritive lesions of the skin and nails to seat itself in the hands and feet will be hereafter noted in connection with the subject of the pains of locomotor ataxy and of those produced by profound mercurial poisoning. And it will be seen in the section on Pathology, that very important conclusions are suggested by the coincidence.

Joined with the burning pains, and the altered skin-nutrition, in the cases of gunshot injury of nerves which we are considering, there is nearly always a marked alteration in the temperature of the parts, either in one direction or the other. In the great majority of instances of ordinary neuralgia after wounds, this alteration is a very considerable reduction of the temperature of the parts supplied by the painful nerves; a change which corresponds with what appears in the vast majority of all cases of division of sensitive nerves, whether pain be set up or not. But, in all examples of the burning pain after injury, Messrs. Mitchell, Moorehouse, and Keen found the temperature of the painful parts notably elevated,

It would appear that there is no form of neuralgia more dreadful, and scarcely any so hopeless, as this burning pain coming on as a sequel to severe nerve injuries. It exercises a profoundly depressing effect upon the whole nervous tone; the most robust men become timid and broken down, and their condition is compared by the American writers to that of hysterical women.

There is another peculiar nutritive affection, first recognized as an occasional consequence of nerve injuries by Messrs. Mitchell, Moorehouse, and Keen, namely, an inflammation of joints, and, although we have no concern here with this symptom, it will be referred to hereafter as throwing interesting light on certain questions of pathology. Certain lesions of secretion will also be specially referred to under the heading of Diagnosis.

II. NEURALGIAS OF INTRA-NERVOUS ORIGIN.—As regards the constitutional conditions with which the several varieties of neuralgia that arise independently of external violence, or disease of extra-nervous tissues, are respectively allied, the following preliminary subdivisions may be made:

---

[*] *Med Times and Gazette*, March 26, 1864.
[†] "London Hosp. Reports," 1866.

1. Neuralgias of malarious origin.
2. Neuralgias of the period of bodily development.
3. Neuralgias of the middle period of life.
4. Neuralgias of the period of bodily decay.
5. Neuralgias associated with anæmia and mal-nutrition.

1. Neuralgias of malarious origin were formerly far more prevalent than they are at present, within the sphere of the English practitioner of medicine; with the general decline of malarial fevers, consequent on improved drainage and cultivation of lands, they have become constantly more scarce. The districts in which they still are found to prevail with any frequency are carefully specified in the interesting report of Dr. Whitley to the Medical Officer of the Privy Council, in the Blue-Book for 1863.

Of course, however, there are a considerable number of persons continually returning to England from countries where malarious diseases are common; and these often bear about with them the effects of paludal poisoning which occasionally exhibits itself in the form of neuralgia. Till very lately, however, I had not happened to come across such cases, although at one time and another I have seen and treated a good many persons returned from India and Africa, whence I judge that neuralgia with this special history is less common than many seem to think. In former times, on the contrary, malarioid neuralgias were so common that they forced themselves on the notice of every practitioner. The term "brow-ague," to this day applied by many medical men to every variety of supraorbital neuralgia, is a relic of the older experience on this point, as is also the very common mistake of expecting all neuralgic affections to present a distinctly rhythmic recurrence of symptoms.

In the year 1864 I published the statement * that, "in a fair sprinkling" of the cases of neuralgia which present themselves in hospital out-patient rooms, ague-poisoning may be suspected; but I was then speaking rather from hearsay than from my own experience, which, in fact, had yielded no clear cases of this sort of neuralgia, and was till just recently unable to reckon up more than two undoubted and one doubtful case of the affection, in all of which the fifth cranial nerve was unattacked. The periodicity in one of the genuine cases was regular tertian, in the other regular quotidian. A semi-algide condition always ushered in the attacks; but this was gradually exchanged, as the pain continued, for a condition in which the pulse was rapid and locomotive, but compressible, and the strength was further depressed. In both these cases there was unilateral flushing of the face, and congestion of the conjunctiva, to a slight degree, during the attack of pain. The pain

---

* "Stimulants and Narcotics," Macmillan, 1854, p. 86.

became duller and more diffused contemporaneously with the lowering of arterial pressure; and, after the disappearance of active pain, moderate tenderness over a considerable tract round the course of the painful nerves remain for some time. There was no distinct development of painful points in the situations described by Valleix; but it should be remarked that the cases were rapidly cured with quinine, which very probably accounts for this circumstance.

Till lately I had not witnessed neuralgia as an after-consequence of tropical malaria-poisoning, although I have had many cases of other diseases, the relics of hot climates, under my care ; but within the last year I have seen a case of extremely severe intercostal neuralgia of a perfectly periodic type occurring in a patient whose constitution had been thoroughly saturated with tropical marsh poison, and in whom the spleen was still much enlarged. The neuralgia was so terrible, and accompanied by such severe algide phenomena at the beginning of the attacks, and such a sense of throbbing as the pain developed, as to lead to serious suspicions of hepatic abscess, for the moment; but the course of events soon corrected this idea.

2. *Neuralgias of the Period of Bodily Development.*—By the "period of bodily development" is here understood the whole time from birth up to the twenty-fifth year, or thereabouts. This is the period during which the organs of vegetative and of the lower animal life are growing and consolidating. The central nervous system is more slow in reaching its fullest development, and the brain especially is many years later in acquiring its maximum of organic consistency and functional power.

That portion of the period of development which precedes puberty is comparatively free from neuralgic affections. At any rate, it is rare to meet in young children with well-defined unilateral neuralgia, except from some very special cause, such as the pressure of tumors, etc. Such neuralgias as do occur are commonly bilateral, and are connected either with the fifth cranial or the occipital nerves.

I must here mention an affection which was quite unknown to my experience, but was brought under my notice by the late Dr. Hillier, who kindly called my attention to the notes of two cases which were published in his interesting work on "Diseases of Children." The cases are those of two female children, aged nine and eleven respectively, in whom the principal symptom was violent and paroxysmal neuralgic headache. In both of these children the existence of cerebral tubercle was suspected, but this proved to be a mistake. In both there were intolerance of light, vomiting, tonic contraction of the muscles of the neck, and occasional double vision; but no impairment of intelligence, no amaurosis, and no paralysis or rigidity of the limbs. Each of these children died rather suddenly, after a violent paroxysm of pain. The main, indeed almost the only characteristic post-mortem change was a marked loss of consistence of tissue, in one case in the pons varolii, in the other in the pons, the medulla oblongata, and the cerebellum. These cases are of the highest possible interest, as are also several other instances of headache in children recorded by Dr. Hillier; notably one in which severe paroxysmal pains were attended with general impairment of brain-power, and, on the occurrence of death from exhaustion, the autopsy reveated an amount of degeneration in the cerebral arteries (as also in the general arterial system) which was astonishing, considering that the child was only ten and a half years old. This case, the full significance and interest of which will be better seen when we come to discuss the subject of pathology, is an example of physical changes in the nervous system, which are usually delayed to an advanced period of life, occurring altogether prematurely, and bringing with them a kind of neuralgic pain which is far more common in the decline than in morning of life. It will be seen presently that functional derangements may be in like manner precociously induced, with the parallel effect of inducing such pains as are ordinarily the product of a later epoch.

From the moment that puberty arrives all is changed in the status of the nervous system. In the stir and tumult which pervade the organism, and especially in the enormous diversion of its nutritive and formative energy to the evolution of the generative organs and the correlative sexual instincts, the delicate apparatus of the nervous system is apt to be overwhelmed, or left behind, in the race of development. Under these circumstances, the tendency to neuralgic affections rapidly increases. It will, however, be seen later that there is a great preponderance of particular varieties of the disease during this time. This period is above all things fruitful in trigeminal neuralgias, especially migraine.

There remains to be noticed the fact that sexual precocity sometimes very much anticipates the peculiar characteristics of the period after puberty. It is well known that in too many instances children are led, by the almost irresistible influence of bad example, to indulge in thoughts and practices which are thoroughly unchildish, and which exercise a powerfully disturbing influence upon the nervous system. A child before the age of puberty ought to be distinguished (if moderately healthy in other respects) by the absence of any tendency to dwell upon his own bodily health. Under the influence of precocious sexual irritation he becomes hypochondriacal and self-centred, and often suffers, not merely from fanciful fears and fanciful pains, but from actual neuralgia, which is sometimes severe. The attacks of migraine which are a frequent affection of delicate children whose puberty occurs at the normal time, are a much earlier torment with children who have early become addicted to bad practices. It is an anticipatory effect upon the constitution, strictly analogous to the production of the so-called "hysteria" in little girls under similar circumstances; and I suppose there is no physician who has not once or twice, at least, met with cases of the latter kind. The existence of any severe neuralgic affection in a young child, if it cannot be traced to tubercle or other recognizable or organic brain-disease is *prima-facie* ground for suspicion of precocious sexual irritation; though, as Dr. Hillier's cases show, it is occasionally produced otherwise. Usually, there are other features which assist in the discovery of precocious sexualism, when it exists; there is a morbid tendency to solitary moping, and a moral change in which untruthfulness is conspicuous.

3. *Neuralgias of the Middle Period of Life.*—By this period is meant the time included between the twenty-fifth and about the fortieth or forty-fifth year. It is the time of life during which the individual is subjected to the most serious pressure from external influences. The men, if poor, are engaged in the absorbing struggle for existence, and for the maintenance of their families; or, if rich and idle, are immersed in dissipation, or haunted by the mental disgust which is generated by *ennui*. The women are going through the exhausting process of child-bearing, and supporting the numerous cares of a poor household, in some cases; or are devoured with anxiety for a certain position in fashionable society for themselves and their children; or again, they are idle and heart-weary, or condemned to an unnatural celibacy. Very often they are both idle and anxious.

It must not be supposed that there is a sharp line of demarcation between this period and the last; nevertheless, there are certain well-marked differences, both in their general tendencies, and as regards the local varieties which are commonest in each. We shall discuss the latter point farther on. At present, it is interesting to remark on the general freedom of the busy middle period of life from first attacks of neuralgia. A person who has had neuralgia previously may, and very likely will, during this epoch, be subject to recurrence of the old affection under stress of exhaustion of any kind. But it is very rare, in my experience, for busy house-mothers or fathers of families to get first attacks of neuralgia during this period of life. It is not the way in which a still vigorous man's nervous system breaks down, if it breaks down at all. Men frequently do break down, of course, at an age when their tissues generally are sound enough, and there is no reason, except on the side of their nervous system, why they should not remain in vigorous health for years. But it is greatly more common for the nervous collapse to take the form of insanity, or hypochondriasis, or paralysis, then that of neuralgia. If a man has escaped the latter disease during the period when the growth of his tissues was active, it is not very often that he falls a victim to it till he begins, physiologically speaking, to grow old.

4. *Neuralgias of Declining Bodily Vigor.*—The period here referred to is that which commences with the first indications of general physical decay, of which the earliest which we can recognize (in persons who are not cut off by special diseases) is perhaps the tendency to atheromatous change in the arteries. The first development of this change varies very considerably in date; but whenever it occurs it is a plain warning that a new set of vital conditions has arisen, and especially notable is its connection with the characters of the neuralgic affections which take their rise after its commencement. The period of declining life is preeminently the time for severe and intractable neuralgias. Comparatively few patients are ever permanently cured who are first attacked with neuralgia after they have entered upon what may be termed the "degenerative" period of existence. I mentioned the existence of commencing arterial degeneration as the special and most trustworthy sign of the initiation of bodily decay; but it is needless to say that this change is often not to be detected in its earliest stage. Something has been done of late years, however, to render its diagnosis more easy. Not to dwell upon the phenomenon of the arcus senilis, which though of a certain value is confessedly only very partially reliable, we may mention the sphygmographic character of the pulse as posessing a real value in deciding the physiological status of the arterial system. There is a well-known form of pulse-curve, square-headed, with marked lengthening of the first or systolic portion of the wave, and with an almost total absence of dicrotism, even when the circulation is rapid, which will often seem to assure us that atheromatous change of the arterial system has commenced, even when the physical characters of inelastic artery are not to be recognized with the finger in any of the superficial vessels by the touch of the finger. Indeed, the latter test is in all cases far less reliable than the sphygmographic trace, except when the arterial change has proceeded to a very marked degree of development.

To a certain extent, the presence or absence of gray hair is of value in deciding whether physiological degeneration has begun. Like the arcus senilis, however, this is only reliable when joined with other indications, for it may be a purely local and separate change, having nothing to do with the general vital status of the body.

5. *Neuralgias which are immediately excited by Anæmia or Mal-nutrition.*—Of the neuralgic affections which can be reckoned in this class, the sole characteristic worthy of note is

the circumstances in which they arise. It would seem that anæmia and mal-nutrition simply aggravate the tendency of existing weak portions of the nervous system to be affected with pain; just as they notoriously do aggravate lurking tendencies to convulsion and spasm. It is very common, for instance, for women to suffer severely from migraine, and other forms of neuralgia, after a confinement in which they have lost much blood. According to my own experience, however, those patients are generally, if not invariably, found to have previously suffered more or less severe neuralgic pain, at some time or other in their history, in the same nerves which now, under the depressing influence of hæmorrhage, have become neuralgic. One of the very worst cases of clavus which I ever saw happened after hæmorrhage in labor; the pain was so severe and prostrating that it appeared likely the patient would become insane. I discovered, on inquiry, that this woman had been liable for many years to headache affecting precisely the same region, on the occasion of any unusual fatigue or excitement.

There is, however, one variety of neuralgia from mal-nutrition which deserves special consideration, viz., that which is occasionally produced as an after-effect of mercurial salivation. I have only seen one instance of this affection, but several are recorded. [Such, at least, is my impression, but I have not been able to find the reports of them.] My patient was a woman of somewhat advanced years when she first came under my notice, but her malady had (though with long intermissions) existed ever since she was a young girl in service. At that early date she was severely salivated by some energetic but misguided practitioner, for an affection which was called pleurisy, but (according to her description) might well have been only pleurodynia, to which servant girls are so very subject. At any rate, the consequences of the medication were most disastrous. Not only did she then and there lose every tooth in her head and suffer extensive exfoliations from the maxillæ, but after this process was over she began to suffer frightfully from neuralgic pains in both arms and in both legs. Tonic medicines and a change to sea-air brought about a tardy and temporary cure; but from that moment her nervous system never recovered itself. Whenever she took cold, or was over-fatigued, or depressed from any bodily or mental cause, she was certain to experience a recurrence of the pains. At the time of her application to me she was suffering from an attack of more than ordinary severity, and which had lasted a long time without showing any signs of yielding. She apparently could not find words to express the acuteness of her sufferings. All along the course of the sciatic nerve in the thigh, all down the course of the middle cutaneous and long saphenous branches of the anterior crural, in the musculo-spiral, radial, and the course of the ulnar nerves, and also, in a more generalized way, in the gastrocnemii, in the soles of the feet, and in the palms of the hands, the pains were of a tearing character, which she described as resembling "iron teeth" tearing the flesh. The pains recurred many times daily; her life was a perfect burden to her, and always had been during these attacks. This patient was under my observation, on various occasions, during several years, and I established the fact that cod-liver oil always did very great good. But it was evident that nothing would remove the tendency to the recurrence of the pains. I should mention, as additional proof of the extent to which the mercurial poison had shattered the nervous system of this woman, that she had violent muscular tremors at the time of her first attack, and on several subsequent occasions. A more completely ruined life was never seen; the poor woman had been on the highway to promotion in the service of a nobleman when she was mercurialized, but her whole prospects were blighted by the serious danger to her health which was caused by the preposterous antiphlogisticism of her medical attendant.

I do not know that the poisonous action of any other metalic poison than mercury has been distinctly shown to produce neuralgic pains of superficial nerves. The action of lead is well known to produce colic, a disease which will be specially dwelt on elsewhere. And undoubtedly a certain amount of aching pain sometimes attends certain stages of lead-palsy of the extensor muscles of the forearm. But I know of no facts pointing to a true saturnine neuralgia. And the chronic poisonous effects of arsenic on the nervous system seem to produce sensory paralysis, rather than pain.

We come now to the consideration of the local varieties of neuralgia. The primary subdivision of them may be made as follows:

I. Superficial Neuralgias. II. Visceral Neuralgias.

I. Of superficial neuralgias a further classification may be made:

(a) Neuralgia of the fifth (trigeminal, or trifacial).
(b) Cervico-occipital neuralgia.
(c) Cervico-brachial neuralgia.
(d) Intercostal neuralgia.
(e) Lumbo-abdominal neuralgia
(f) Crural neuralgia.
(g) Sciatic neuralgia.

This arrangement is that of Valleix, and appears to me substantially correct.

(a) The most important group of neuralgias are those of the fifth cranial nerve.

Neuralgia of the fifth nerve always exhibits itself in the especial violence in certain foci, which Valleix was the first to define with accuracy. These foci are always in points where the nerve becomes more superficial, either in turning out of a bony canal, or in penetrating fasciæ. In the ophthalmic division of the nerve the following possible foci are noticeable: (1) The supra-orbital, at the notch of that name, or a little higher, in the course of the frontal nerve; (2) the palpebral, in the upper eyelid; (3) the nasal, at the point of emergence of the long nasal branch, at the junction of the nasal bone with the cartilage; (4) the ocular, a somewhat indefinite focus within the globe of the eye; (5) the trochlear, at the inner angle of the orbit.

In the superior maxillary division the following foci may be found: (1) The infra-orbital, corresponding to the emergence of the nerve of that name from its bony canal; (2) the malar, on the most prominent portion of the malar bone; (3) a vague and indeterminate focus, sonewhere on the line of the gums of the upper jaw; (4) the superior labial, a vague and not often important focus; (5) the palatine point, rarely observed, but occasionally the seat of intolerable pain.

In the inferior maxillary division the foci are: (1) The temporal, a point on the auriculo-temporal branch, a little in front of the ear; (2) the inferior dental point, opposite the emergence of the nerve of that name; (3) the lingual point, not a common one, on the side of the tongue; (4) the inferior labial point, only rarely met with.

Besides these foci in relation with distinct branches of the trigeminus, there is one of especial frequency which corresponds to the inosculation of various branches. This is the parietal point, situated a little above the parietal eminence. It is small in size—the point of the little finger would cover it. It is the commonest focus of all.

Neuralgia may attack any one, or all, of the three divisions of the nerve; the latter event is comparatively rare. Valleix, indeed, holds a different opinion; but this seems to me to arise from the fact that his definition of neuralgia was too narrow to include a large number of the milder cases of neuralgia, which are, nevertheless I believe, decidedly of the same essential character with the severer affections. The most frequent occurrence is the limitation of the pain to the ophthalmic division, and incomparably the most frequent foci of pain are the supra-orbital and the parietal.

The most common variety of trigeminal neuralgia is migraine, or sick-headache, as it is often called. This is an affection which is entirely independent of digestive disturbances, in its primary origin, though it may be aggravated by their occurrence. It almost always first attacks individuals at some time during the period of bodily development. Under the influences proper to this vital epoch, and often of a further debility produced by a premature straining of the mental powers, the patient begins to suffer headache after any unusual fatigue or excitement, sometimes without any distinct cause of this kind. The unilateral character of this pain is not always detected at first; but, as the attacks increase in frequency and severity, it becomes obvious that the pain is limited to the supra-orbital and its twigs, with sometimes also the ocular branches. In rare cases, as in all forms of nèuralgia, the nerves of both sides may be affected; I have already observed that this seems to be relatively more common in young children. If the pain lasts for any considerable length of time, nausea, and at length vomiting, are induced. This is followed at the moment by an increase in the severity of the pain, apparently from the shock of the mechanical effect; but from this point the violence of the affection begins to subside, and the patient usually falls asleep. The history of the attacks negatives the idea that the vomiting is ordinarily remedial. This symptom merely indicates the lowest point of nervous depression; but it may happen that a quantity of food which has been injudiciously taken, lying as it does undigested in the stomach, may of itself greatly aggravate the neuralgia, by irritation transmitted to the medulla oblongata. In such a case vomiting may directly relieve the nerve-pain. When the patient awakes from sleep, the active pain is gone. But it is a common occurrence—indeed it always happens when the neuralgia has lasted a long time—that a tender condition of the superficial parts remains for some hours, perhaps for a day or two. This tenderness is usually somewhat diffused, and not limited with accuracy to the foci of greatest pain during the attacks.

Sick headache is not uncommonly ushered in by sighings, yawning, and shuddering—symptoms which remind us of the prodromata of certain graver neuroses, to which, as we shall hereafter see, it is probably related by hereditary descent. In its severer forms, migraine is a terrible infliction; the pain gradually spreads to every twig of the ophthalmic division; the eye of the affected side is deeply bloodshot, and streams with tears; the eyelid droops, or jerks convulsively; the sight is clouded, or even fails almost altogether for the time, and the darts of agony which shoot up to the vertex seem as if the head were being split down with an axe. The patient cannot bear the least glimmer of light, nor the least motion, but lies quite helpless, intensely chilly and depressed, the pulse at first slow, small and wiry, afterward more rapid and larger, but very compressible. The feet are generally actually, as well as subjectively, cold. Very often, toward the end of the attack, there is a large excretion of pale, limpid urine.

Another variety of trigeminal neuralgia which infests the period of bodily development is that known as clavus hystericus: clavus, from the fact that the pain is at once severe, and limited to one or two small definite points, as if a nail or nails were being driven into the skull. These points correspond either to the supra-orbital or the parietal, or, as often happens, to both at once. But for the greater limitation of the area of pain in clavus, that affection would have little to distinguish it from migraine, for the former is also accompanied with nausea and vomiting when the pain continues long enough; and in both instances it is obvious that there is a reflex irritation propagated from the painful nerve. The adjective hystericus is an improper and inadequate definition of the circumstances under which clavus arises. The truth is, that the subjects of it are chiefly females who are passing through the trying period of bodily development; but there is no evidence to show that uterine disorders give any special bias toward this complaint. Both migraine and clavus are often met with in persons who have long passed their youth; but their first attacks have nearly always occurred during the period of development.

One circumstance in connection with well-marked clavus appears worth noting, as somewhat differentiating it from migraine. It is, I think, decidedly more frequently the immediate consequence of anæmia than they; but it does not appear, from my experience, that the chlorotic form of anæmia is any more provocative of it than is anæmia from any other cause. Some of the worst cases of clavus, probably, that have ever been seen were developed in the old days of phlebotomy. It was then very common for a delicate girl, on complaint of some stitch of neuralgia or muscular pain in the side, to be immediately bled to a large extent, with the idea of checking an imaginary commencing pleurisy. The treatment, so far from curing the pain and the dyspepsia (which it produced), often aggravated them; whereupon the signs of inflammation were thought to be still more manifest, and more blood was taken. Under such circumstances the most complete anæmia was developed, and very often the patient became a martyr to clavus in its severest forms. One does not now very frequently meet with the victims of such mistaken practice; but I have seen one [since writing this I have seen another case (*vide* cardiac neuralgia, *infra*)] very severe case of clavus produced by loss of blood (in a subject who was doubtless predisposed to neuralgic affections, to judge from his family history). The case was that of a boy who accidentally divided his radial.

The middle period of life is not, according to my experience, fruitful in first attacks of trigeminal neuralgia. But, when the neuralgic tendency has once declared itself, there are many circumstances of middle adult life which tend to recall it. Over-exertion of the mind is one of the most frequent causes, especially when this is accompanied by anxiety and worry; indeed, the latter has a worse influence than the former. In women, the exhaustion of hæmorrhageal parturition, or of menorrhagia, and also the depression proauced by over-suckling, are frequent causes of the recurrence of a migraine or clavus to which the patient had been subject when young. The middle period of life is very obnoxious to severe mental shocks, which are more injurious than in youth, because of the diminished elasticity of mind which now exists; and the same may be said of the influence of severe bodily accident of a kind to inflict damage on the central nervous system. Special mention ought to be made, in the case of women, of the disturbing influence of the series of changes which close the middle portion of their life, viz., the involution of the sexual organs. It would seem as if every evil impression which has ever been made on the nervous system hastens to revive, with all its disastrous effects, at this crisis. Latent tendencies to facial neuralgia are particularly apt to reassert their existence, and they are usually accompanied and aggravated by a tendency to vaso-motor disturbance, which not unfrequently seems to be the most distressing part of the malady. I have several times been consulted by women undergoing the "change," whose chief complaint was of disagreeable flushings and chills, especially of the face; and, on inquiring further, one has found that they were suffering from severe facial neuralgia, which, however, alarmed and distressed them less than did the vaso-motor disturbance, and the giddiness, etc., which were an evident consequence of it.

It is, however, the final or degenerative period of life which produces the most formidable varieties of facial neuralgia. Neuralgia of the fifth, which have previously attacked an individual, may recur at this time of life without any special character, except a certain increase of severity and obstinacy. But trigeminal neuralgias, which now appear for the first time, are usually intensely severe, and nearly or quite incurable. These cases correspond with the affection named by Trousseau tic epileptiforme, and it is of them, doubtless, that Romberg is speaking, when he says that the true neuralgias of the fifth rarely occur before the fortieth year of life. These neuralgias are distinguished by the intense severity of the pain, the lightning-like suddenness of its onset, and the almost total impossibility of effecting more than a temporary palliation of the symptoms. But they are also distinguished by another circumstance which too often escapes attention, namely, they are almost invariably connected with a strong family taint of insanity, and very often with strong melancholy and suicidal tendencies in the patient himself, which do not depend on, and are not commensurate with, the severity of the pain which he suffers. It may seem a strong view to take, but I must say that I regard a well-developed and typical neuralgia, of the type we are now speaking of, as an affection in which the mental centres are almost as deeply involved as in the fifth nerve itself; though, whether this is an original part of the disease, or a mere reflex effect of the affection of the trigeminal nerve, I am not prepared to say. Other reflex affections are common enough in this kind of facial neuralgia, and especially spasmodic contractions of the facial muscles, which, indeed, often form one of the most striking features of the malady, the attacks of pain being accompanied by hideous involuntary grimaces. Even in the earlier stages of the disease there is usually some degree of the same thing, as, for instance, spasmodic winking. In the great majority of cases, after a little time, exquisitely tender points are formed in the chief foci of pain; in the intervals between the spasms the least pressure on these points is sufficient to cause agony, and a mere breath of wind impinging on them will often reproduce the spasm. Yet, in the height of the acute paroxysm itself, the patient will often frantically rub these very parts in the vain attempt to produce ease; and it has often been noticed that such friction has completely rubbed off the hair or whisker on the affected side: this happens the more easily, because the neuralgic affection itself impairs the nutrition of the hair and makes it more brittle, as we shall have occasion to show more fully hereafter. The general appearance of a confirmed neuralgic of the type now described is very distressing, and the history of his case fully corresponds to it. He is moody and depressed, he dreads the least movement, and the least current of air; he hardly dares masticate food at all, more especially if the inferior maxillary division of the nerve be implicated (as is generally the case sooner or later), for this movement re-excites the pain with great violence. Nutrition is very commonly kept up by slops, and is thus very insufficiently maintained: this failure of nutrition is itself a decidedly powerful influence in aggravating the disease. And there is a still further calamity which is not unlikely to occur. The patient may fly to the stupefaction of drink as a relief to his sufferings, and, if he has once experienced the temporary comfort of drunken anæsthesia, is excessively likely to repeat the experiment. But this is another and one of the most fatally certain methods of hastening degeneration of nerve-centres, and the ultimate effect, therefore, is disastrous in every way.

Although the neuralgias of the degenerative period are thus fatally progressive, on the whole, there are some curious occasional anomalies. Many cases are recorded, and I have myself seen such, in which the attacks of pain, after reaching a very considerable degree of intensity, have ceased for many months, whether under the influence of remedies or not it is difficult to say with certainty, but probably far more from independent causes. Whatever may be the reason of these sudden arrests, however, certain it is that they are very seldom permanent, the pain returning sooner or later, like an inexorable fate.

(b) **Cervico-occipital Neuralgia.**—As Valleix has remarked, there are several nerves (in fact, the posterior branches of all the first four spinal pairs) which are more or less frequently the seat of this affection. But among them all there is none comparable to the great occipital, which arises from the second spinal pair, for the frequency and importance of its neuralgic affections. This nerve sends branches to the whole occipital and the posterior parietal region. On the other hand, the second and third spinal nerves help to make up the superficial cervical branch of the cervical plexus which is distributed to the triangle between the jaw, the median line of the neck, and the edge of the sterno-mastoid, and those to the lower part of the cheek. Then there is the auricular branch, which starts from the same two pairs, and supplies the face, the parotid region, and the back of the external ear. Then the small occipital, distributed to the ear and to the occiput. And, finally, superficial descending branches of the plexus. These, altogether, are the nerves which at various points, where they become more superficial, form the foci of cervico-occipital neuralgia.

The most typical example of this form of neuralgia which has fallen under my notice occurred (after exposure to cold wind) in a lady about sixty years of age, who had all her life been subject to neuralgic headache approaching the type of migraine, and who came of a family in which insanity, apoplexy, and other grave neuroses, had been frequent. The pain centred very decidedly in a focus corresponding to the occipital triangle of the neck; it recurred at irregular intervals, and in very severe paroxysms, lasting about a minute. It was interesting to follow the history of this case in one respect. It afforded a clear illustration of the manner in which local tenderness is developed; for during the first three or four days the patient, so far from complaining that the painful part was tender on pressure, experienced decided relief from pressure, although she experienced none from mere rest, however carefully the neck might be supported. But in the course of a few days an intensely painful spot developed itself in the occipital triangle, and the back of the ear became excessively tender. All manner of remedies had been tried in this case, without the slightest success and especially there was a large amount of speculative medication, on the theory of the probably "rheumatic" or "gouty" nature of the affection. Nothing was doing the least good to the pain, and meantime the old lady's digestion and general health and spirits were suffering very severely. Blistering was now suggested, and the affection yielded at once. The relief afforded must have been very complete, to judge by the warm gratitude which the patient expressed. The subsequent history of this patient illustrates several points which will engage our attention under the section of Pathology. It may be just mentioned here, that she suffered, twelve months later, from a hemiplegic attack of paralysis.

The tendency of cervico-occipital neuralgias is to spread toward the lower portions of the face, as observed by Valleix; in this case they become, sometimes, undistinguishable from neuralgias of the third division of the trigeminus. In the early stages of the disease, if the physician had been lucky enough to witness them, the true place of the origin of the pain would have been easily recognizable; at a later date it sometimes needs great care, and a very strict interrogation of the patient, to discover the true history of the disease. Sometimes, even, a cervico-occipital neuralgia which spreads in this way causes great irritation and swelling of the submaxillary and cervical glands; and I have known a case of this kind mistaken for commencing glandular abscess. The pain and tension were so great in this case, and the constitutional disturbance was so considerable, that the presence of deep-seated pus was strongly suspected, and the propriety of an incision (which would have been a hazardous proceeding) was seriously canvassed.

Experience is too limited, to judge by what I have personally seen, and the recorded cases with which I am acquainted, to enable us to say anything with confidence of the conditions, as to age and general nutrition of the body, which specially favor the occurrence of cervico-occipital neuralgia. Apparently, however, there is much reason for thinking that the immediately exciting cause of it is most frequently external cold. I have known it produced several times in the same person, by sitting in a draught which blew strongly on the back of the neck. And I am inclined to think that it is seldom the first form of neuralgia which attacks a patient, but usually occurs in those who have previously suffered from neuralgic pains either of the trigeminus or of some other superficial nerve. I have known it once to occur in a person, thus predisposed to neuralgic affections, in consequence of reflex irritation from a carious tooth, as was proved by its cessation on the extraction of the latter, although there was no facial pain.

(c) *Cervico-brachial Neuralgia.*—This group includes all the neuralgias which occur in nerves originating from the brachial plexus, or from the posterior branches of the four lower cervical nerves. The most important characteristic of the neuralgias of the upper extremity is the frequency, indeed almost constancy, with which they invade, simultaneously or successively, several of the nerves which are derived from the lower cervical pairs. The neuralgic affections of the small posterior branches (distributed to the skin of the lower and back part of the neck) are comparatively of small importance. But the "solidarite," which Valleix so well remarked, between the various branches of the brachial plexus, causes the neuralgias of the shoulder, arm, forearm, and hand to be extremely troublesome and severe, owing to the numerous foci of pain which usually exist. Perhaps Valleix's description of these foci is somewhat over-fanciful and minute; but the following among them which he mentions I have repeatedly identified; (1) An axillary point, corresponding to the brachial plexus itself; (2) a scapular point, corresponding to the angle of the scapula. (It is difficult to identify the peccant nerve here; the one to which it apparently corresponds, and to which Valleix refers it, is the subscapular; but we are accustomed to think of this as a motor nerve. Still, it is certain that pressure on a painful point existing here will often cause acute pain in the nerves of the arm and forearm.) (3) A shoulder point, which corresponds to the emergence, through the deltoid muscle, of the cutaneous filets of the circumflex; (4) a median-cephalic point, at the bend of the elbow, where a branch of the musculo-cutaneous nerve lies immediately behind the median-cephalic vein; (5) an external humeral point, about three inches above the elbow, on the outer side, corresponding to the emergence of the cutaneous branches which the musculo-spiral nerve gives off as it lies in the groove of the humerus; (6) a superior ulnar point, corresponding to the course of the ulnar nerve between the olecranon and the epitrochlea; (7) an inferior ulnar point, where the ulnar nerve passes in front of the annular ligament of the wrist; (8) a radial point, marking the place where the radial nerve becomes superficial, at the lower and external aspect of the forearm. Besides these foci, there are sometimes, but more rarely, painful points developed by the side of the lower cervical vertebræ, corresponding to the posterior branches of the lower cervical pairs.

The most common seat of cervico-brachial neuralgia has been, in my experience, the ulnar nerve, the superior and inferior points above mentioned being the foci of greatest intensity; an axillary point has also been developed in one or two cases which I have seen. Rarely, however, does the neuralgia remain limited to the ulnar nerve; in the majority of cases it soon spreads to other nerves which emanate from the brachial plexus. A very common seat of neuralgia is also the shoulder, the affected nerves being the cutaneous branches of the circumflex. I am inclined to think, also, that affections of the musculo-spiral, and of the radial near the wrist, are rather common, and have found them very obstinate and difficult to deal with. One case has recently been under my care in which the foci of greatest intensity of the pain were an external humeral and a radial point; but besides these there was an exquisitely painful scapular point. In another case the pain commenced in an external humeral and a radial point, but subsequently the shoulder branches of the circumflex became involved. A most plentiful crop of herpes was an intercurrent phenomenon in this case, or rather, was plainly dependent on the same cause which produced the neuralgia.

Median cephalic neuralgia is an affection which used to be comparatively common in the days when phlebotomy was in fashion, the nerves being occasionally wounded in the operation. I have only seen it in connection with this cause, that is to say, as an independent affection. One such case has been under my care. But a slight degree of it is not uncommon, as a secondary symptom, in neuralgia affecting other nerves. The traumatic form is excessively obstinate and intractable.

In the neuralgias of the arm we begin to recognize the etiological characteristic which distinguishes most of the neuralgic affections of the limbs, namely, the frequency with which they are aggravated, and especially with which they are kept up and revived when apparently dying out, the muscular movements. In the case above referred to, of neuralgia of the subscapular, musculo-spiral (cutaneous branches), and radial, the act of playing on the piano for half an hour immediately revived the pains, in their fullest force, when convalescence had apparently been almost established.

There is a special cause of cervico-brachial neuralgias which is of more importance than, till quite lately, has ever been recognized, namely, reflex irritation from diseased teeth. The subject of these reflex affections from carious teeth has been specially brought forward by Mr. James Salter, in a very able and interesting paper in the "Guy's Hospital Reports" for 1867; and Mr. Salter informs me that he has been surprised by the number of cases of reflex affections, both paralytic and neuralgic, of the cervico-brachial nerves, produced by this kind of irritation, and that he agrees with me in thinking that a peculiar organization or disposition of the spinal centres of these nerves must be assumed in order to account for the fact.

The liability of particular nerves in the upper extremity to neuralgia from external injuries requires a few words. The nerve which is probably most exposed to this is the ulnar. Blows on what is vulgarly called the funny-bone are not uncommon exciting causes of neuralgia in predisposed persons, and cutting wounds of the ulnar a little above the wrist are rather frequent causes. The deltoid branches of the circumflex and the humeral cutaneous branches of the musculo-spiral are much exposed to bruises and to cutting wounds. So far as I know, it is only when a nerve trunk of some size has been wounded that neuralgia is a probable result. Wounds of the small nervous branches in the fingers, for instance, are very seldom followed by neuralgia. I have no statistics to guide me as to the effect of long-continued irritation applied to one of these small peripheral branches, but it is probable that that might be more capable of inducing neuralgia. As far as my own experience goes, however, it would appear that a more common result is convulsion of some kind, from reflex irritation of the cord.

(*d*) *Dorso-intercostal Neuralgia.*—This is one of the commonest varieties of neuralgia, and yet it is very likely to be confounded with other affections not neuralgic in their nature. The disorder with which it is especially liable to be confounded is myalgia, which will be fully described in another chapter, and which, when developed in the region of the body to which we are now· referring, is commonly spoken of as pleurodynia, or lumbago (according as it affects the muscles of the back or of the side), or muscular rheumatism. It must be owned that the severer forms of this affection can scarcely be distinguished from true intercostal neuralgia by anything in the character or situation of the pains. It will be seen, hereafter, however, that myalgia has its own specific history, which is very characteristic ; at present, it is sufficient to remember that it is often extremely like neuralgia when situated in the dorso-intercostal region.

Dorso-intercostal neuralgia is an affection of certain of the dorsal nerves. These nerves divide, immediately after their emergence from the intervertebral foramina, into an interior and a posterior branch. The latter sends filaments which pierce the muscles to be distributed to the skin of the back; the former, which are the intercostal nerves, follow the intercostal spaces. Immediately after their commencement they communicate with the corresponding ganglia of the sympathetic. Proceeding outward, they at first lie between two layers of intercostal muscles, and, after giving off branches to the latter, give off their large superficial branch. In the case of the seventh, eighth and ninth intercostal nerves, which are those most liable to intercostal neuralgia, the superficial branch is given off about midway between the spine and the sternum. The final point of division, at which superficial filets come off, in all the eight lower intercostal nerves, is nearer to the sternum; and is progressively nearer to the latter in each successive space downward. There are thus, as Valleix observes, three points of division: (1) At the intervertebral foramen ; (2) midway in the intercostal space; (3) near to the sternum. And there are three sets of branches (reckoning the posterior division) which respectively make their way to the surface near to these points.

In one of its forms, intercostal neuralgia is one of the commonest of all neuralgic affections. I refer to the pain beneath the left mamma, which women with neuralgic tendencies so often experience, chiefly in consequence of over-suckling, but also from exhaustion caused by menorrhagia or leucorrhœa, and especially from the concurrence of one of the latter affections with excessive lactation. It is especially necessary, however, to guard against mistaking for this affection a mere myalgic state of the intercostal or pectoral muscles, which often arises in similar circumstances with the addition of excessive or too long continued exertions of these muscles. "Hysteric" tenderness also sometimes bears a considerable resemblance, superficially, to true intercostal neuralgia, in cases where the genuine disease does not exist.

A less common but very remarkable variety of intercostal neuralgia than that just mentioned, is the kind of pain which attends a good many cases of herpes zoster, or shingles. It is only of recent years that any essential connection between zoster and neuralgia has been suspected. The occurrence of neuralgia as a sequel to zoster had indeed been mentioned by Rayer, Recamier, and Piorry, but the essential nature of the connection between the two diseases was evidently not suspected by Lecadre, when, as late as 1855, he published his valuable essay on intercostal neuralgia. M. Notta was one of the first to present connected observations on the subject. But it was much more fully discussed in a paper published by M. Barensprung, in 1861. [*Ann. der Charite-Krakenhauser zer Berlin, ix.*, 2, p. 40. *Brit. and For. Med. Rev.*, January, 1862.] This author showed the absolute universality with which unilateral herpes, wherever developed, closely followed the course of some superficial sensory nerve, and gave reasons, which will be discussed hereafter, for supposing that the disease originates in the ganglia of the posterior roots, and that the irritation spreads thence to the posterior roots in the cord, causing reflex neuralgia. We shall have more to say on this matter. Meantime, it seems to be established, by multiplied researches, that, though unilateral herpes may and often does occur without neuralgia, and neuralgia without herpes, the concurrence of the two is due to a mere extension of the original disease which is a nervous one.

In young persons, zoster is not attended with severe neuralgia, but a curious half-paretic condition of the skin, in which numbness is mixed with formication, or with a sensation as of boiling water under the skin, precedes the outbreak of the eruption by some hours, or by a day or two. Painless herpes is commonest in youth. I remember, for instance, that, in an attack of shingles which I suffered about the age of eleven, there was at no stage any acute pain; only, in the pre-eruptive period, for a short time, I had the curious sensations referred to above: and the same thing has occurred in all the patients below puberty that I have seen, if they complained at all. From the age of puberty to the end of life, the tendency of herpes to be complicated with neuralgia becomes progressively stronger. The course of events varies much in different cases, however. In adult and later life the symptoms usually commence with a more or less violent attack of neuralgic pain, which is succeeded, and generally, though not always, displaced by the herpetic eruption. The latter runs its course, and after its dis appearance the neuralgia may return, or not. In old people it almost always does return, and often with distressing severity and pertinacity. Six weeks or two months is a very common period for it to last, and in some aged persons it has been known to fix itself permanently, and cease only with life. In these subjects a further complication sometimes occurs. The herpetic vesicles leave obstinate and painful ulcers behind them, which refuse to heal, and which worry the patient frightfully, the merest breath of air upon them sufficing to produce agonizing darts of neuralgic pain. I have known one patient, a woman over seventy years of age, absolutely killed by the exhaustion produced by protracted suffering of this kind.

The foci of pain in intercostal neuralgia are always found in one or more of the points, already enumerated, at which sensory nerves become superficial. In long-standing cases acutely tender points are developed in one or more of these situations; not unfrequently the most decided of these spots is where it gets overlooked, namely, opposite the intervertebral foramen. H. G., a young woman aged twenty-six, who applied to me at Westminster Hospital, had suffered for twelve months from an irregularly intermitting but very severe neuralgia at the level of the seventh intercostal space of the left side. The violence of the pain was sometimes excessive, and when the paroxysm lasted longer than usual it generally produced faintness and vomiting. This patient had no sign of tenderness anywhere in the anterior or lateral regions, though the pain seemed to gird round the left half of the chest as with an iron chain, but an exquisitely tender spot, as large as a shilling, was found close to the spine; pressure on this always induced a strong feeling of nausea.

As an illustration of the herpetic variety of dorso-intercostal neuralgia, running a severe but not protracted course, I may relate the case of a medical man whom I formerly attended. This gentleman was about thirty-two years of age, and a highly neurotic subject: inter alia, he had already suffered from a severe and protracted sciatica; and, very shortly before the herpetic attack, had been jaundiced from purely nervous causes. His nervous maladies were undoubtedly caused by over-brain-work. In this case the neuralgia developed itself during the latter half of the eruptive period, which was rather unusually lengthened. It occupied the seventh, eighth, and ninth intercostal spaces of the side affected with herpes, and was very violent and acute, so that the patient expressed himself as almost "cut in two" with it. The pain ceased even before the vesicles had perfectly healed; a rather unusual occurrence in my experience. I shall refer to this case hereafter, as an example of what I believe to be the effect of a particular method of treatment in lessening the tendency to after-neuralgia. The result of my experience is certainly this—that if a case of herpes in an adult, or still more in an aged person, be left to itself, the amount of after-neuralgia will very closely correspond with the severity of the eruptive symptoms.

There is a variety of intercostal neuralgia which is of more importance than the commoner kinds. Occurring mostly in persons who have passed the middle age, it possesses the characters of obstinacy and severity which belong to the neuralgias of the period of bodily decay. It is at first unattended with any special cardiac disturbance. By-and-by, however, it begins to attract more careful attention from the fact that the severer paroxysms extend into the nerves of the brachial plexus of the affected side, so that pain is felt down the arm. In the midst of a paroxysm of intercostal and brachial pain, it may happen that the patient is suddenly seized with an inexpressible and deadly feeling of cardiac oppression, and, in fact, the symptoms of angina pectoris, such as they will be described in a future chapter, become developed. A case of this kind is at present under my care at the Westminster Hospital. The patient is a man only fifty-six years of age, but whose extreme intemperance has produced an amount of general degeneration of his tissues such as is rarely seen except in the very aged; he has the most rigid radial arteries, and the largest arcus senilis, I think, that I ever saw. This man has long been subject to attacks of violent intercostal neuralgia, and a recent access assumed the type of unmistakable angina. It is very probable that his coronary arteries have now become involved in the degenerative process. In this case, before the development of any marked anginal symptoms, the paroxysmal pain, from being merely intercostal, had come to extend itself into the left shoulder and arm.

Intercostal neuralgia not unfrequently accompanies, and is sometimes a valuable indication of, phthisis. I do not mean to say that the vague pains in the chest-walls, which are so very common in phthisis, are to be indiscriminately accounted neuralgia; on the contrary, they are, in the large majority of instances, merely myalgic, and arise from the participation of the pectorals, or intercostals, or both, in the mal-nutrition which prevails in the organism generally. But it happens, sometimes that a distinctly intermitting neuralgia occurs as an early symptom of phthisis ; in fact, where there is a predisposition to neurotic affections, I believe that this is not very uncommon. The subjects are generally women ; they are mostly of that class of phthisical patients who have a quick intelligence, fine soft hair, and a sanguine temperament. I have had one male patient under my care: this was a young gentleman aged eighteen, in whom a neuralgic access came on with so much severity, and caused so much constitutional disturbance, that the idea of pleurisy was strongly suggested. The paroxysms returned at irregular intervals for a considerable period: they were quite unlike myalgic pains, not only in their character, but more especially with respect to the circumstances which were found to provoke their recurrence. They were the first symptoms which lead to any careful examination of the chest; it was then found that there were prolonged expiration and slight dulness, at one apex. At this period, wasting had not seriously commenced; but, on the other hand, there was an extraordinary degree of debility for so early a stage of phthisis. I am inclined to think that self-abuse was the principal cause both of the phthisis and the neuralgia, acting doubtless on a predisposed organism, for his family was rather specially beset with tendencies to consumption. I may add here, that it has appeared to me that young persons with phthisical tendencies are specially liable to neuralgic affections as a consequence of self-abuse.

A special variety of intercostal neuralgia is that which attacks the female breast. The nerves of the mammæ are the anterior and middle cutaneous branches of the intercostals; and they are not unfrequently affected with neuralgia, which is sometimes very severe and intractable. Dr. Inman has very properly pointed out that a large number of the cases of so-called "hysterical breast" are really myalgic, and are directly traceable to the specific causes of myalgia; but there is no question in my mind that true neuralgia of the breast does occur, and indeed is frequent, relatively to the frequency of neuralgias generally. There are several kinds of circumstances under which it is apt to occur. In highly-neurotic patients it may come on with the first development of the breasts at puberty; and it may be added that this is especially apt to occur where puberty has been previously induced by the unfortunate and mischievous influences to which we had occasion to refer in speaking of certain other neuralgiæ. A neuralgia of the left breast occurred in a patient of mine, who attended the Westminster Hospital. She was only twelve years of age, and small of stature, but the mammæ were considerably developed. The face was haggard, there was an almost choreic fidgetiness about the child, and a very unprepossessing expression of countenance; the result of inquiries left no doubt that the patient was much addicted to self-abuse; and it seemed probable that to this was due the fact that menstruation had come on, and was actually menorrhagic in amount.

A very painful kind of mammary neuralgia is experienced by some women during pregnancy; but more commonly the mammary pains felt at this period are mere throbbings, not markedly intermittent in character, and plainly dependent on mechanical distention of the breast: such affections are not to be reckoned among true neuralgiæ. A true neuralgia of a very severe character is sometimes provoked by the irritation of cracked nipples. I have seen a delicate lady, of highly-neurotic temperament, and liable to facial neuralgia, most violently affected in this way. Vain attempts had been made for several consecutive days to suckle the infant from the chapped breast; when suddenly the most severe dorso-intercostal neuralgia set in. The attacks lasted only a few seconds each, but they recurred almost regularly every hour, and were attended with intense prostration, and sometimes with vomiting. Discontinuance of suckling was found necessary, for even the application of the child to the sound breast now sufficed to arouse a paroxysm of pain. Complete rest, protection of the breast from air and friction, and the hypodermic injection of morphia, rapidly relieved the sufferer.

(e) *Dorso-lumbar Neuralgia.*—The superficial branches of the spinal nerves emanating from the lumbar plexus are considerably less liable to be affected with severe and well-marked neuralgia than are the dorso-intercostal nerves. Pains in the abdominal walls, which are a good deal like neuralgia, are not uncommon; but the majority of them will be found, on careful observation, to be myalgia. At least, this has been the case in my own experience.

When true neuralgia of the superficial branches of the lumbo-abdominal nerves occurs, it develops itself in one or more of the following foci: (1) Vertebral points, corresponding to the posterior branches of the respective nerves; (2) an iliac point, about the middle of the crista ilii; (3) an abdominal point, in the hypogastric region; (4) an inguinal point, in the groin, near the issue of the spermatic cord, whence the pain radiates along the latter; (5) a scrotal or labial point, situated in the scrotum or in the labium majus.

Such is the description given by Valleix; for my own part, I cannot say that I have seen enough cases to test its accuracy. I believe it to be generally correct, yet it may fairly be doubted whether the author might not have revised his description had the natural history of myalgic affections been as carefully investigated as it has since been. The hypogastric foci of pain of which he speaks are at least open to considerable suspicion, as it will be shown, in the chapter on Myalgia, that an extremely common variety of the latter affection is situated in this region, and the severity of the pain which it often produces might well cause it to be mistaken for a genuine neuralgia.

I have, however, seen three or four cases in which the very complete intermittence of the paroxysms, without any perceptible relation to the question of muscular fatigue, left no doubt in my mind of the really neuralgic character of the malady. In one of these instances, oddly enough, the exciting cause appeared to be fright; and this was as severe a case as one often sees. The patient was a woman of middle age, and much depressed by the long continuance of a profuse leucorrhœa. As she was walking along the street, a herd of cattle, in a somewhat irritable and disorderly condition, came suddenly toward her; she immediately began to suffer pain just above the crest of the ilium, and at the lumber region, and, most acutely, in the labium majus of one side; and then pain returned daily, about 10 A. M., lasting for half an hour with great severity. This woman's family history was remarkable : her mother had been paraplegic, her sister was a confirmed epileptic, and two of her children had suffered from chorea.

In two other cases of lumbo-abdominal neuralgia which were under my care, there were also very painful points in the spermatic cord and in the testicle. One of these cases will be referred to under the head of Visceral Neuralgia. Another case, in which severe quasi-neuralgic pain was referred to the groin, will be described in the chapter on the Pains of Hypochondriasis.

(*f*) *Crural Neuralgia.*—This appears to be rare as an independent affection occurring primarily in the crural nerve. Valleix had only seen it twice in all his large experience, and I have never seen it myself. Neuralgic pain of the crural nerve is almost always a secondary affection arising in the course of a neuralgia, which first shows itself in the external pudic branch of the sacral plexus; or else occurring as a complication of sciatica. A remarkably severe example of the latter occurrence was observed in an old man who still occasionally attends the Westminster Hospital. He has been a martyr to the most inveterate bilateral sciatica for between two and three years; and, within the last three months, it has extended itself into the cutaneous branches of the curval nerves of both thighs. So great an aggravation of the pain is produced by any muscular movement, that the patient can only walk at the slowest possible pace, moving each foot forward only a few inches at a time. The bilateral distribution of the pain is remarkable in this case; but there can be no doubt of its really neuralgic character, from the truly intermittent way in which it recurs, and the absence of any history whatever to point in the direction of rheumatism, gout, or syphilis.

The nervous supply to the skin of the anterior and external portion of the thigh includes : (1) The middle cutaneous, (2) the internal cutaneous, and (3) the long saphenous branch of the anterior crural nerve ; (4) the cutaneous branch of the obturator; and (5) the external cutaneous nerve, derived from the loop formed between the second and third lumbar nerve. The sensitive twigs derived from the two latter sources, equally with the branches of the anterior crural, are liable to be secondarily affected by neuralgia, which commences in the lumbo-abdominal nerves ; but it must be a rare event for them to be the seat of a primary neuralgia. The only occasion on which I have seen anything which looked like the latter was in the case of a porter, who, in straining to lift a very heavy load, ruptured some part of the attachment of the tensor vaginæ femoris. But the susceptibility of all the nerves of the front of the thigh to secondary or reflex neuralgia receives numerous illustrations. The extremely severe pain at the internal aspect of the knee-joint, which is such a common symptom in morbus coxæ, is evidently a reflex neuralgia of the long saphenous nerve, the ultimate irritation being situated in the branches of the obturator nerve which supply the hip-joints. For some reason unexplained, it happens that this saphenous nerve is specially liable to be affected in a reflex manner: for instance, this happens in a considerable number of cases of sciatica. I have a lady now under my observation, in whom the secondary neuralgia of the saphenous nerve has become even more intolerable than the pain in the sciatic, which was the nerve primarily affected. The pain in these cases very frequently runs down the inner and anterior surface of the leg to the internal ankle. Sometimes the branches of the anterior crural become the seat of intensely painful points in the course of a long-persisting sciatica. A patient at present under my care has a spot, about the size of a shilling, just at the emergence of the middle cutaneous branch from the fascia lata, which is intensely and persistently tender to the touch, and the skin here is so exquisitely sensitive to the continuous galvanic current that the application of moistened sponge-conductors, with a current of only fifteen Daniell's cells, causes intolerable burning pain; whereas at every other part of the limb the current from twenty-five cells can be borne without much inconvenience.

(*g*) *Femoro-popliteal Neuralgia, or Sciatica,*—This is one of the most numerous and important groups of neuralgia; but, notwithstanding that there are plenty of opportunities for studying it, I venture to think it is very commonly mistaken for different and non-neuralgic diseases, and they for it. The rules of diagnosis which will be laid down for all the neuralgiæ would nevertheless prevent these errors, if carefully attended to.

Sciatica is a disease from which youth is comparatively exempt. Valleix had collected one hundred and twenty-four cases, and in not one was the patient below the age of seventeen, only four were below twenty. In the next decade there were twenty-two; in the next, thirty; and the largest number of cases, thirty-five, occurred between the ages of forty and fifty. This completely tallies with my own experience, and appears to afford some support to a suspicion I have formed, that the chief exciting cause of sciatica is the pressure exercised on the nerve in locomotion, and that this cause exercises its maximum influence when the period of bodily degeneration commences. It is further remarkable that, in elderly persons (whose habits of locomotion are of course more limited), the proportion of fresh cases rapidly diminishes; and also that above the age of thirty the number of male patients greatly exceeds that of female patients attacked. All this seems to point in the same direction.

According to my observation, there are three distinct varieties of sciatica. The first of these is obscure in its origin, but may be said, in general terms, to be connected with a nervous temperament of the highly impressible kind, which is more or less like what we call "hysteric," not only in the female, but also in male patients. The subjects of this kind of sciatica are mostly young persons, and hardly ever more than middle-aged; they are generally found to be liable to other forms of neuralgia; and the actual attack of sciatica is produced by some fatigue or mental distress, which at other times might have brought on sick headache, or intracostal neuralgia, etc. Very many of these patients are anæmic; and chlorotic anæmia seems specially to favor the occurrence of the affection. The greater number of the victims are females, and in very many, whether as cause or effect, there is impeded, or at least imperfect, menstruation. This kind of sciatic pain is not usually of the highest degree of intensity, but it generally spreads into a great many branches, both in a direct and a reflex manner. It is probable that this variety of the disease is, at least very often, dependent upon, or much aggravated by, an excited condition of the sexual organs; certainly, I have observed it with special frequency in women who have remained single long after the marriageable age, and in several male patients there has been either the certainty or a strong suspicion of venereal excess. Sciatica of this kind also occurred in the case of a single woman aged about thirty, who to my knowledge was excessively addicted to self-abuse.

The second variety of sciatica occurs for the most part in middle-aged or old persons who have long been subject to excessive muscular exertion, or have been much exposed to damp and cold, or who have been subject to the combined influence of both these kinds of evil influence. One must also include, I think, in this group a considerable number of cases where the age is not so advanced, but the patient has been obliged, by the nature of his business, to maintain the sitting posture daily, for hours together, exercising pressure on the nerve; this is especially liable to happen in these persons.

The sufferers from this variety of sciatica are mostly, as already said, of middle age or more; but this statement must be understood to be made in the comparative sense, which refers rather to the vital status of the individual than to the mere lapse of years. Many of these people have hair which is prematurely gray, and in some the existence of rigid arteries, together with arcus senilis, completes the picture of organic involution, or senile degeneration. In particular cases, where depressing influences have been at work for a long time, or unusually active, these appearances rectify the false impression we should otherwise derive from learning the mere nominal age of the person; this is especially often the case with regard to patients who have for a long time drunk to excess. The prematurely and permanently gray hair (it will be seen hereafter that permanency of grayness is an important point), together with well-marked inelasticity of arteries, very often tells a tale which is most useful in informing us, not only of the vital status of the patient, but of the kind of sciatica under which he labors; and also influences our prognosis seriously. There is otherwise a somewhat deceptive air about the appearance of many of these degenerative cases; for instance, a ruddy complexion is not uncommon, nor the retention of considerable, or even great, muscular strength. It is probable that these appearances deceived Valleix and many others,.or they could hardly have failed, as they have, to observe the frequency of the degenerative type among the most numerous group of sciatic patients, namely, those between thirty and fifty years of age. These persons are not truly "robust," although at a hasty glance they might at first seem to be so. It would be a serious mistake to omit the search for the important vital evidences which have been referred to, since these therapeutic and prognostic indications are of the highest value.

A prominent feature in this kind of sciatica is its great obstinacy and intractability, Another, equally marked, is the tendency to the development of spots around the foci of severest pain which are intensely and permanently tender, and the slightest pressure on which is sufficient to set up acute pain, This is a symptom much less developed, if developed at all, in the variety of sciatica which we first discussed. The places which are especially apt to present this phenomenon of tenderness are as follows : (1) A series, or line of points, representing the cutaneous emergence of the posterior branches, which reaches from the lower end of the sacrum up to the crista ilii ; (2) a point opposite the emergence of the great and small sciatic nerves from the pelvis ; (3) a point opposite the cutaneous emergence of the ascending branches of the small sciatic, which run up toward the crista ilii; (4) several points at the posterior aspect of the thigh, corresponding to the cutaneous emergence of the filets of the crural branch; (5) a fibular point, at the head of the fibula, corresponding to the division of the external popliteal; (6) an external malleolar, behind the outer ankle; (7) an internal malleolar.

I have already mentioned that in sciatica the pain frequently spreads in a reflex manner to nerves which are connected, by their origin from the plexus, with the sciatic. It will be remembered, also, that I related cases in which the formation of tender points, in the course of the nerves thus secondarily affected, was even more distinct and remarkable than anywhere in the branches of the sciatic itself.

Another circumstance which distinguishes the form of sciatica which we are now describing is, the degree in which (above all other forms of neuralgia) it involves paralysis of motion. [The subject of the complication of neuralgia will be treated in a general manner farther on; but it seems necessary to note here the special liability of sciatic patients to this and to the most material complications]. By far the largest part of the motor nervous supply for the whole lower limb passes through the trunk of the great sciatic; it might therefore be naturally expected that a strong affection of the sensory portion of the nerve would produce, in a reflex manner, some powerful effect upon the motor element. This effect is most frequently in the direction of paralysis. Complete palsy is rare, but in a large proportion of cases which have lasted some time there will be found, independently of any wasting of muscles, a positive and considerable loss of motor power. It is of course necessary to avoid the fallacy which might be produced by neglecting to observe whether movement was restricted merely in consequence of its painfulness. Not long since, I had occasion to test the electric sensibility in a case of sciatica, in which there was extremely severe pain, affecting chiefly the peroneal region of the leg, and great weakness of the leg, amounting to inability for walking. The gastrocnemius could hardly be got to contract at all, when the most powerful Faradic current was directed upon the nerve in the popliteal space of the affected limb, though the muscle of the sound side reacted with great vigor.

Anæsthesia is also a common complication of sciatica, far commoner, I venture to think, than it has been represented either by Valleix, or Notta. It is necessary, however, to be explicit on this point. In the early stages, both of this form of sciatica, and of the milder variety previously described, there is almost always partial numbness of the skin previous to the first outbreak of the neuralgic pain, and during the intervals between the attacks. By degrees this is exchanged, in the milder form, for a generally diffused tenderness around the foci of neuralgic pain, while other portions of the limb remain more or less anæsthetic. In the severer forms it sometimes happens that, besides an intense tenderness of the skin over the painful foci, there is diffused tenderness over the greater part or the whole of the surface of the limb. But it is important to remark that both in the anæsthetic and the hyperæsthetic conditions (so called) the tactile sensibility is very much diminished. I have made a great many examinations of painful limbs, in sciatica, and have never failed to find (with the compass points) that the power of distinctive perception was decidedly lowered.

Convulsive movements of muscles are met with in a moderate proportion of cases of sciatica in middle and advanced life, in which affection they are entirely involuntary. They differ from certain spasmodic movements not unfrequently observed in the milder form (and especially in hysteric women), for these are more connected with morbid volition, and are, in truth, not perfectly involuntary. In several cases of inveterate sciatica I have seen violent spasmodic flexures of the leg upon ~~upon~~ the thigh. Cramps of particular muscles are occasionally met with. I have seen the flexors of the toes of the affected limb violently cramped, and in one case there was agonizing cramp of the gastrocnemius. It is chiefly at night, and especially when the patient is falling asleep, that this kind of affection is apt to occur.

A third variety of sciatica is the rather uncommon one, so far as my experience goes, in which inflammation of the tissues around the nerve is the primary affection, and the neuralgia is mere secondary effect, from mechanical pressure on the nerve, which, however, is not apparently itself inflamed. I believe that these cases are sometimes caused by syphilis, and sometimes by rheumatism. One of the most violent attacks of sciatic pain which ever came under my notice was in a syphilized subject, a discharged soldier, who had been the victim of severe tertiary affections, and had been mercilessly salivated into the bargain. This unfortunate man suffered dreadful agony, which was aggravated every night, but was never totally absent. The pain started from a point not far behind the great trochanter: pressure here caused intolerable darts of pain, which ramified into every offshoot of the sciatic nerve, as it seemed, and made the man quite faint and sick. Large doses of iodide of potassium, together with the prolonged use of cod-liver oil, completely removed the pain and tenderness. It need hardly be said that cases of this kind are essentially different, and require perfectly different principles of treatment, from neuralgias in which the disturbance originates within the nervous tissues themselves.

The chronic rheumatism does also, occasionally, affect the sheath of the nerve in such a manner as to produce a deposit which sets up neuralgic pain, must also be admitted, although I believe the number of such cases to be preposterously over-estimated by careless observers. It has several times happened that a patient has come under my care with so-called "rheumatic affection of the nerves" of the thigh and leg, and that on examination one has found all the symptoms and clinical history of a neurosis, but not the slightest valid argument for a diagnosis of the rheumatic diathesis. Indeed, upon this point, I think it is time that a decided opinion should be expressed. I firmly believe that a large number of sciatic patients have their health ruined by treatment directed to a supposed rheumatic taint which is purely imaginary. The state of medical reasoning, suggested by the way in which too many practitioners decide that such and such pains are rheumatic in their origin, is a melancholy subject for reflection. Nearly always it will be found, on cross-examination, that the state of the urine has been made the basis of a confident diagnosis; the practitioner will tell you that the urine was loaded, *i, e.*, with lithtaes. He ignores the fact that nothing is more common, in neurotic patients who are perfectly guiltless of rheumatic propensities, than a fluctuation between lithiasis and oxaluria, neither of which phenomena, under the circumstances, indicates any more than a temporary defect of secondary assimilation of food, produced by nervous commotion. I may perhaps find room, on a future page, for a few further remarks on the subject; at present I only put in a caution against too ready an acceptance of the rheumatic hypothesis.

II. VISCERAL NEURALGIAS.—*Uterine and Ovarian Neuralgia.* This is an important group of neuralgic affections, and one which I cannot help thinking is strangely misappreciated, very often, in a therapeutic point of view. In one aspect these affections possess a special interest, namely this, that they are more frequently dependent on peripheral irritation for their immediate causation than any other group of neuralgias. If we consider the great copiousness of the nervous supply to the uterus and ovaries, and the powerfully disturbing character of the functional processes which are periodically occurring in these organs, we shall be at no loss to understand how this may be. The amount force of the peripheral influence and which are brought to bear upon the central nervous system by the functions of the uterus and ovaries are greater than any that emanate from the diseases and functional disturbances of any other organ in the body.

The most common variety of peri-uterine neuralgia is that which attends certain kinds of difficult menstruation. It would be hardly correct to give the name of neuralgia to the pain existing in these very numerous cases of dysmenorrhœa in which the suffering is apparently altogether dependent on the mere retention or difficult escape of the menstrual fluid, although the character of the pain often resembles the neuralgic type. There is another group of dysmenorrhœal affections however, in which the pain may fairly be called neuralgic, since it is apparently independent of the circumstances of the discharge of menstrual fluid, and simply attends the process, seemingly on account of a naturally-exaggerated irritability of the organs concerned. There is a large class of young women in whom, and more especially before marriage, the time of menstruation is always marked by the occurrence of more or less severe pain. Formerly I used to believe that this pain was relieved on the occurrence of the discharge, but I have seen too

4

many cases of a contrary nature to retain this opinion. I now believe that the subjects of the kind of menstrual pain to which I am referring are naturally endowed with a very irritable nervous apparatus of the pelvic organs, and that there is a certain character at once of immaturity and excitability in their sexual organs, especially in the virgin condition. So far from these females being disposed to sterility, as is too often the case with those dysmenorrhœal subjects whose troubles depend upon occlusion, distortion, or narrowing of the outlets, they are often extremely apt to the generative function; and, what is more, the full and natural exercise of the sexual function appears necessary to the health of their organs, as is shown by the fact that these menstrual pains lose their abnormal character, completely or in great part, after marriage, and especially after child-bearing. The contrast between the two types of dysmenorrhœal patients is sharply brought out by the two following cases:

CASE I.—S. M., a housemaid, aged twenty-three when first under my notice, was the picture of physical health and strength, very intelligent, and a girl of excellent character and most industrious habits. At every menstrual period, however, she suffered, for some hours previously to the occurrence of the flow, from severe pain in the uterine region, which was tumefied and tender. Hot hip-baths gave some relief, apparently by hastening the discharge; as soon as the latter was established, the pain rapidly subsided. This young women married a healthy and vigorous young man, but has never had any children, and at the date of my last inquiries still suffered periodically from her old troubles.

CASE II.—Mrs. B. was married at the age of twenty-six. Up to the date of her marriage she used to suffer the most severe pain at every menstrual period, the pain; however, bore no relation to the freedom of the discharge, but always lasted about the same length of time, under any circumstances, or was only less or more according as the general bodily vigor was greater or less at the moment. From the date of marriage these troubles steadily declined; a child was born at the end of twelve months, and the menstrual troubles have never resumed a serious shape up to the present time, a period of nearly nine years. This lady is herself a neuralgic subject, liable to migraine in circumstances of fatigue, and suffering horribly from it during her pregnancies; and she comes of a family in whom the nervous temperament is strongly developed.

It must not always be concluded, because the menstrual pain is very severe before the discharge and is relieved at or soon after its appearance, that the case is one of occlusion, and not of neuralgia. There is a class of cases in which the affection appears to be a very severe ovarian neuralgia, attended with a vaso-motor paralysis which causes great engorgement of the ovary and consequent difficult of "ovulation." I have seen several instances which I could not explain in any other way.

CASE III.—One patient I particularly remember, from the fact that she was always attacked with dreadful pain, which was sometimes seated in one groin and sometimes in the other, but was regularly attended with large and palpable tumefaction of the ovary, which began to subside when the discharge commenced. This woman married rather late, but her mensural troubles immediately became less, and she became pregnant and was happily delivered, nearly as soon as was possible. She, too, was a decidedly neuralgic subject, independently of her tendency to dysmenorrhœal ovarian pain.

In some women who remain single long after the marriageable age, ovarian or uterine neuralgia becomes a constantly-recurring torment, not only at the menstrual period, but at various other times when they are depressed or fatigued in body or mind. As might be expected, this tendency is greatly aggravated in the rarer cases where the patient's mind dwells in a conscious manner on sexual matters, especially if by an evil chance she becomes addicted to self-abuse. Among the many reproaches that have been thrown upon the indiscriminate use of the speculum in examining unmarried women, it has often been urged that it tends to excite sexual feelings. I do not for a moment doubt that this is the case, or that the indiscriminate use of the instrument is altogether indefensible. But I expect that neuralgic pain of the uterus or ovaries, in unmarried women, connected with an already irritable condition of the sexual organs, has often been the reason why such women have applied for advice and has consequently been examined with the speculum; and that the same thing has frequently happened in the case of women who have been left widows at a time of life when the sexual powers were still in full vigor. These patients deserve great pity.

The peripheral irritation which gives rise to peri-uterine neuralgia is not always originally seated in the organs of generation. The following are various sources of external irritation which I have known to produce the affection :

1. Ascarides in the rectum sometimes produce pelvic neuralgia. A woman, aged thirty-four, single, was under my care in King's College Hospital many years ago, under suspicions of ulcerated cervix. On examination, no lesion could be detected. It was discovered that the rectum was infested with ascarides, and, after the use of appropriate vermifuges and tonics, the patient entirely lost the uterine pains and also a tormenting pruritus vaginæ, from which she suffered. This woman had at various times suffered from neuralgic headache a good deal.

2. Profuse and intractable leucorrhœa, whether associated or not with ulceration of the cervix, may produce peri-uterine

neuralgia, even of great severity, when there are strongly-marked neurotic tendencies. It must be noted, however, that many cases of pain in leucorrhœal subjects, which superficially bear the aspect of neuralgia, turn out on closer investigation to be merely examples of myalgia of the abdominal muscles or aponeuroses.

3. Calculus in the kidney, or in the ureter, sometimes causes intolerable ovarian neuralgia. In the case of a woman who was under my care at the Chelsea Dispensary, some years ago, this was the unsuspected origin of severe neuralgic pains in the left ovary, which recurred several times a day, and which certainly contributed to the patient's death by the exhaustion which they produced.- A calculus was found tightly impacted in the uterer, near the kidney.

4. Prolapsus uteri sometimes gives rise to severe peri-uterine neuralgia, or what appears to be such; though it is difficult here to draw the line between neuralgia and myalgia. The commonest kind of pains from prolapsus uteri are not neuralgic in their nature at all, but are of a "bearing down" character, and probably depend upon actual contractile movement of the walls of the uterus.

5. The presence of tumors, either cancerous or fibroid, in the uterus or its appendages, gives rise, frequently, to severe and indeed almost intolerable pains of a distinctly intermittent character. In the early stages of cancerous diseases these pains are usually felt at the lower part of the back ; in the later stages they are felt also in the hypogastric region, and are then much more severe.

6. Ulcer of the cervix, of a non-malignant kind, probably sometimes gives rise to neuralgic pain of the uterus, though this is not so severe as in cancer.

7. Large masses of scybalous fæces, impacted in the rectum, will occasionally, by the pressure which they exert on nerves, set up violent neuralgia of uterus or ovaries, the true nature of which is accidentally discovered by the use of aperients which unload the intestine and put an end to the suffering. No doubt it is chiefly in persons with neuralgic predisposition that this effect is produced; for, common as is the occurrence of extreme constipation in women, it is comparatively very rare for us to hear of distinctly neuralgic pain being caused by it.

8. The condition known as "irritable uterus," ever since Gooch's classical description of it, is always attended with uterine pain, which is continuous, but is liable to periodical exacerbations of great severity. In this disorder there is no recognizable physical disease of the pelvic organs, and the patient will generally be found to have suffered neuralgia in other parts of the body on previous occasions. [There is some difference of opinion about this affection : some authors (*e.g.*, Hanfield Jones) considering it as distinct from the true neuralgias.]

9. Reflex irritation, the source of which is in some quite distant part of the body, has in many recorded instances occasioned uterine neuralgia, in highly-predisposed persons. I have seen one case in which severe pain of this kind was clearly proved to have been excited by the presence of a carious tooth which was itself little, if at all, painful, but the removal of which at once cured the pelvic pain.

Neuralgia of the urethra is an affection which is occasionally seen, both in males and females. I have observed it three times; all these cases were apparently traceable to the effects of excessive self-abuse. The male subject was an unmarried man, aged forty-two, of cadaverous appearance, much emaciated, with clammy, perspiring skin, and habitual coldness of the extremities; he suffered much from dyspepsia and palpitation of the heart. The pain ran along the under side of the penis, which was very large, with an elongated prepuce. The paroxysms were severe, and came on chiefly in the morning, soon after he awoke. No remedies did this man any permanent good, and he passed out of my sight, being at that time in a condition of wretched febleness, and with symptoms of threatened dementia. Of the female subjects, one was a married woman, who accused her husband of impotence, and from her account it would certainly appear that effective connection had never taken place; the hymen was completely destroyed, however. The neuralgic pains recurred nightly in several paroxysms, and were especially severe about the time of the monthly periods. In this case the patient was, she stated, induced to give up her malpractices; at any rate, the pain subsided in a manner which could not be well accounted for by any direct influence of the medicinal treatment. The other female patient was a widow in whom the morbid habit was suspected from her general appearance, and from the existence of enlarged clitoris and other signs of irritation about the external parts: she became rather rapidly phthisical, and suffered severely from neuralgic headaches.

Neuralgia of the bladder has been specially described by various writers: the pain is usually spoken of as seated at the neck of the bladder, and as accompanied by frequent desire to micturate. I have seen two cases, both in women: the first was eventually discovered to be an instance of malignant disease of the fundus of the bladder; the other was apparently the result of a long-continued menorrhœal flux, which had greatly impaired the health, and produced extreme anæmia. In neither of these instances was the pain referred to the external meatus, as in the female patients above mentioned who were suffering from urethral neuralgia. I have never seen the extreme examples of vesical neuralgia described by some writers, in which actual paralysis of the coats of the bladder was secondarily produced; but the reflex influence of the neuralgic affection in both the examples just mentioned appeared to produce great weakening of the muscular power of the rectum, occasioning most obstinate and troublesome constipation.

It would appear, from recorded cases, that both the bladder and the uterus are liable to be affected with neuralgia from malarious influences; but I have never chanced to see any such cases.

Neuralgia of the kidney is spoken of by several writers, and I suppose there there is no doubt that it may exist as a special neurotic disease with obvious organic cause. For my own part, I cannot say that I have ever seen it except in instances where there was either the certainty, or a very strong suspicion, that the cause was the mechanical pressure and irritation of a calculus within the kidney. The diagnosis of the simple functional disorder must be excessively perplexing; for in the first place there is the greatest difficulty in making sure that the pain is not external, and seated either in the muscles of the back, or in the superficial dorsal or lumbar nerves, and certainly I am strongly inclined to suspect that this has been really the case in many examples of so-called renal neuralgia. That neuralgia of the kidney may arise secondarily, as a reflex extension of pelvic neuralgia, does, however, appear probable enough; for it is almost certain that in the latter affection at least, the vaso-motor nerves of the kidneys must be strongly influenced in a reflex manner; since the crisis or acme of a paroxysm of pelvic pain is not unfrequently attended with a copious secretion of pale urine.

Neuralgia of the rectum has been carefully described by Mr. Ashton, but is probably not often seen except by practitioners who possess special opportunities of observing rectal diseases. In the one pure case which has fallen under my notice the patient complained of acute paroxysmal cutting pains extending about one inch within the anus, and, as these were greatly increased by defecation I suspected the existence of fissure. Nothing of the kind, however, was found on examination; and the pain ultimately yielded to repeated subcutaneous injections of atropine. This patient had got wet through, and had sat in his damp clothes, getting thoroughly chilled; the pain came on with great suddenness and severity, and the tenderness which has been mentioned was developed very quickly. probably the influence of cold and wet is among the commonest causes of the complaint. Mr. Ashton also reckons as causes, reflex irritation from other parts of the alimentary canal, and the influence of malaria. He observes that the subjects of the affection are most frequently anæmic, and of a generally excitable and deranged susceptibility, and that females, who, from menorrhagia, or frequent child-bearing with much hæmorrhage, have lost a great deal of blood, are specially predisposed.

Neuralgia of the testis (as an independent affection and not a mere extension of lumbo-abdominal neuralgia) is fortunately a much less common malady than the corresponding affection of the ovary; as might indeed be expected, from the much less degree of functional perturbation to which, in ordinary physiological circumstances, the former organ is exposed than the latter. Except from actual growths within the testis, of which it was a mere symptom, I have never seen neuralgia of the testis save from one of three causes. In one remarkable example it was produced as a reflex effect of severe herpes preputialis. Secondly, it is sometimes observed as a symptom of calculus descending the ureter. And, thirdly, I have seen it several times undoubtedly produced by excessive self abuse.

The occurrence of testicular neuralgia, in one case of epilepsy, as to the cause of which I had been previously much puzzled, led to the discovery of the real origin of the fits. I should observe here that I do not believe that self-abuse is ever more than an immediately exciting cause of epilepsy, a predisposition to the disease having previously existed in all cases. In the patient just referred to, there was a family history of epilepsy, but it was difficult to explain the exciting cause until this was suggested by the occurrence of neuralgic pain in the testicle. The patient relinquished his habit, and both the pain and the epilepsy ceased, and, for some twelve months during which I had him under observation, had not recurred at all. A medical friend has informed me of an instance in which the same habit had produced a neuralgia of the testis so severe as to strongly tempt the patient to castrate himself, and he would probably have done so but that he was too much of a coward with regard to physical pain. The attacks of pain were so severe as frequently to produce vomiting and the greatest prostration.

Hepatic Neuralgia.—It must be allowed that the evidence even for the existence of neuralgia of the liver is at present in an unsatisfactory state. At the same time, there are carefully-recorded cases, by Trousseau and other * writers of unquestionable authority, which leave no doubt in my mind, corroborated as they are by a certain amount of experience of my own, that such a form of neuralgia really exists. I must, of course, be understood to refer to something altogether different from the spasmodic pain which is produced by the difficult passage of a gall-stone toward the bowel. I have now seen several cases in which, as it appeared to me, there was sufficient evidence of neuralgic pain seated in the liver itself, and not dependent either on gall-stone or any so-called organic diseases of the viscus.

---

* Trousseau, Clinique Medicale. Vanlair, " Des dieffrentes Formes du Nevralgies," Journ de Med. de Bruxelles, tome xl.

The subjects of hepatalgia are probably never troubled only by pain in the liver; they are persons of a nervous temperament, in whom a slight shock to, or fatigue of, the nervous system, habitually provokes neuralgic attacks; the pain localizing itself sometimes in the branches of the trigeminal, sometimes in those of the sciatic, sometimes in the intercostal nerves, etc. In one instance which has been under my observation, the attacks of hepatalgia alternated with cardiac neuralgia assuming the type of a rather severe angina pectoris. In another case the patient, a man aged sixty-seven, was very liable to attacks of intermittent abdominal agony, in which one could hardly doubt that the pain was located in the colon, and was attended with paralytic distention of the bowel; the peculiar feature of the case being the sudden way in which the symptoms would appear and depart, independently of any recognizable provocation or the use of any remedies. On two separate occasions this patient was attacked with pain of a precisely similar kind, but limited to the right hypochondrium, attended with great depression of spirits, and followed by a well-pronounced jaundice. So remarkable was the conjunction of symptoms in these two attacks that a strong suspicion of biliary calculus was raised, but not the slightest confirmation of this idea could be obtained; and indeed one symptom—vomiting—which nearly always attends the painful passage of a biliary calculus, was altogether absent.

Putting aside a considerable number of cases in which "pain in the liver" was vaguely complained of by patients who were plainly hypochondriacal, and whose account of their own sufferings could not be relied on, I have altogether seen five instances of what I regard as genuine hepatalgia. The first of these was very remarkable in its history and in all its features. The patient was a respectable girl of eighteen, subject to migraine, who had reason to fear that she had become pregnant, though this proved, ultimately, not to be the case. Under these circumstances she was attacked with intermittent pains, in the right hypochondrium, of intolerable severity; resembling, in fact, the pain of biliary calculus, but without the sense of abdominal constriction, and without any vomiting. These recurred daily at about the the same hour in the morning, for about ten days; when rather suddenly, a jaundiced tint appeared upon the face, and very shortly the whole skin was colored bright yellow; there was intense mental apathy; the urine was loaded with bile-pigment, and the fæces clay-colored. This state of things lasted only about a week, and then very rapidly disappeared; but as the jaundice subsided there was a partial recurrence of the neuralgic pains, which, for a day or two, were as severe as they had ever been. The other four cases of hepatalgia which I have seen, including that of the man above mentioned, have all been in persons in advanced life; but, except the latter, neither of them displayed any symptoms of disordered biliary secretion; and the diagnosis (as to situation, for the character of the attacks was manifestly neuralgic) rested mainly on the fact that the pain radiated to the shoulder.

There remains to be noticed one clinical feature of the disease, which, I believe, is characteristic; namely, the peculiar mental depression which attended all the cases I have seen, but was most marked in the two in which jaundice occurred. In the girl above referred to, the apathy, during the period when there was jaundice but no pain, was even alarming; it reminded one of the mental state in commencing catalepsy; during the painful stages it was more like the gloom of suicidal melancholia. Of course, the acute mental anxiety which this patient had suffered would account for a good deal of this; but the symptom was as distinct, though less severe, in the case of an elderly lady, whom I have attended on another occasion for migraine; here there was no recognizable source of anxiety; and, on the other hand, there was no reason to suspect the retention of bile-elements in the blood. It seems, therefore, as if an essentially depressing influence on the mind was excited by hepatic neuralgia; or else, that emotional causes are the chief source of the malady.

Neuralgia of the Heart.—If there be any hesitation in treating this disease as exactly conterminous with angina pectoris, it can, I think, be only reasonably justified on two grounds: In the first place, it may be urged that acute pain of the neuralgic type is not always present in angina pectoris; and, secondly, it may be urged that many cases of painful neurosis of the heart have been observed, in which the recurrence of pain with some amount of cardiac embarrassment has gone on for years, whereas the popular conception of true angina almost necessarily involves rapid fatality.

There is doubtless some force in these objections, especially in the second, for it does seem rather inconvenient to call by the same name so deadly a disorder as the worst form of angina, and so comparatively harmless a malady as some of those instances of chronic tendency to spasmodic pain of the heart which are not very uncommon, and in which the patient survives, perhaps, to an old age. Yet, after all, there is the greatest difficulty in drawing any rational line of distinction; for the basis of the affection seems the same in every case, whether pain or spasm be the predominant feature, and whether the course of the disease be long or short. All that appears to be necessary for its production is a certain originally neurotic temperament (with possibly some congenital weakness or some post-natal disease of that part of the spinal-cord centres which Von Bezold has described as furnishing three-fourths of the propulsive power of the heart) and the presence of almost any kind of difficulty or embarrassment of the action of the heart. The most common source of this embarrassment is perhaps failure of nutrition in the muscular walls of the heart, from disease of the coronary arteries. Indeed, it is not known that any organic change of the heart or great vessels, even of the slightest kind, is necessary to the production of angina; on the contrary, there is every reason to think that mere fatigue and depression may bring on the attacks in persons of a strongly nervous temperament. For my own part, I am inclined to believe, however that there really always is disease somewhere in the cardiac centre of the spinal cord, though that disease may consist in no more than a disposition to minute interstitial atrophy. But we shall say more about this presently.

It is at any rate certain that cardiac neuralgia is always a most grave complaint, from the almost total uncertainty whether succeeding attacks will not involve a fatal amount of spasm. As for the expression angina pectoris, it is just one of those mischievous terms which, arising out of the mystified ignorance in which the elder physicians found themselves as to the pathology of internal diseases, have since been attached in turn to various definite organic changes, with none of which they had any essential connection; and it is therefore much to be wished that it could be altogether done away with. At the same time, there is so much that is peculiar in the case of cardiac neuralgia, owing to the importance of the organ affected, that it will be necessary here to treat not merely its symptoms, but also its diagnosis, prognosis, etiology, pathology, and treatment, in a separate and continuous manner.

Clinical History and Symptoms.—Cardiac neuralgia usually shows itself for the first time with considerable abruptness. The patient may or may not have been consciously ill before the actual seizure, but it rarely happens, even when the heart has notoriously been the subject of some organic disease, that there has been any thing to lead him to expect the kind of attack from which he now suffers. In the midst of some little unusual effort, or even without this kind of provocation, suddenly the patient is attacked with severe pain, usually at the lower part of the sternum; this pain darts through to the back and left shoulder, and nearly always runs down the left arm. Sometimes, indeed, it is felt acutely over a large area of the chest, and runs down both arms; this is the case in a patient now under my care, in whom the affection is more obviously a neurosis, and less attended with coarse organic changes, than is usually the case. Along with the pain, which is always very distressing, but varies greatly in severity in different cases, there is a variable amount of another sensation which can be compared to nothing but cramp, or rather compression; the patient usually describes it as feeling as if some one were grasping the heart in his hands, and, when this sensation is at all prominent, the idea of impending death is most strongly impressed on the sufferer's mind. His outward appearance seems to confirm the idea. In cases where the sense of compression is great, the face is of an ashen gray; the lips white, with a faint livid tinge; the pulse small, feeble, and unrhythmical, or imperceptible, at the wrist; cold perspiration breaks out upon the face; in short, all the signs of approaching dissolution are present. In cases where the suffering is chiefly or entirely confined to severe pain, of a darting or burning character, the state of the circulation is often different. The heart bounds against the ribs, in rapid and painful palpitation, the face is flushed deep crimson, the pulse at the wrist is large, bounding, but very compressible; in fact, the outward appearance of the patient is so different from that of one who suffers from the more depressing kind of angina, that it is difficult to consider the two affections as essentially similar. But there can be no question, if we carefully examine the matter, that they are mere varieties of the same disorder, especially as they both may successively occur in the same person.

The course of cardiac neuralgia varies extremely. Supposing the malady to be purely neurotic, and not complicated with organic disease, which forms a constant source of cardiac embarrassment, then the patient may only experience one or two attacks, under some special circumstances of exhaustion, which may never recur; or, on the other hand, he may develop a strong tendency to cardiac neuralgia which may beset him during almost any number of years. In the latter case, it is an even chance whether the patient will at last sink from the anginal affection; for, even supposing him to escape any fatal intercurrent disease of an independent nature, the fatal event may be at last produced by cerebral softening, or by apoplexy, or other central nervous disease. In fact, the frequency with which the latter kind of termination occurs is very significant of the essential nature of the disease.

The manner in which cardiac neuralgia commences varies very greatly. In the celebrated case of Dr. Arnold, the first attack did not occur till he was forty-seven years of age; it at once assumed full intensity, and proved fatal in two hours and a half. There is also reason to believe that Dr. Arnold's father died in a first attack of angina. I have myself known a first attack prove fatal in the course of an hour; there was very considerable ossification of the coronary arteries and fatty degeneration of the heart-walls. Again, there are many cases which commence gradually, and with great mildness, and with little appearance of danger to life in the first attacks; but the subsequent attacks are progressively more severe and dangerous up to a fatal result, after weeks, months, or years. On the other hand, I have known three instances in which the first attacks of spasmodic heart-pain very nearly proved fatal, but the subsequent fits were milder (in one there was no second attack): all those patients are living, six, eight, and three years respectively, after their first attacks.

It can hardly be doubted that neuralgic spasm is the true cause of sudden death in some cases of stenosis of the aortic orifice, which, but for some accidental circumstances, would not have died suddenly at all, but would have gone through a long and gradual course of deterioration. I particularly remember an instance in which extreme and calcareous constriction of the aortic orifice, in a boy not yet come to puberty, was entirely unsuspected, until one day, in running fast, he screamed out and fell down, and was almost instantaneously dead. I remember another case very similar, in which extreme mitral constriction produced almost as sudden death, apparently from painful spasm, under the same kind of exertion. On the other hand, sudden death, when produced by the form of heart-disease which (as Dr. Walshe points out) is most likely to cause such a catastrophe, viz., aortic regurgitation pure, without hypertrophy, does not seem to be due to painful spasm, but to simple and complete failure of the muscular power, and is perhaps partly of the nature of paralysis from a syncopal condition of the brain, the unhypertrophied heart having become for the moment unable to supply blood enough to the brain to carry on nervous function at all.

A good instance of the form which angina takes, when the element of organic cardiac change is well pronounced, was afforded by the case of a young gentleman recently under my care. He was twenty-one years of age, and from early boyhood had been accustomed to a great deal of muscular exercise; in fact, it is probable that he had undermined his health by the frequent and extraordinarily long walks which he took, for his frame was particularly small and slight, and the muscles small and soft. He came of a family in whom the tendency to neurotic disorders is obviously very strong; both his father and his brother are subject to bad attacks of migraine, and he had himself repeatedly suffered from the same thing. The family disposition, altogether, is highly nervous and excitable. The remarkable circumstance in this young gentleman's case is, that although he had taken for years an extraordinary amount of pedestrian exercise (including mountain-climbing), and latterly had exchanged this for the even more trying exertion of rowing, he had never suffered from any noticeable symptom of cardiac distress up to the very day of his anginal attack. For some months, however, he had been growing thin and pale, and I had given him certain cautions, and had made him take cod-liver oil and steel, as I entertained some fears of his becoming phthisical. On the day of the attack there was nothing particular in his appearance, but he complained of a slight cold, and had no appetite for his six o'clock dinner. He retired to rest at eleven o'clock, having taken a small dose of laudanum and chloric ether for his cold. In less than half an hour he awoke out of his sleep in fearful agony; so severe and prostrating was the anginoid pain that he had the greatest difficulty in crawling out of bed to unlock his door. I found him bathed in cold sweat, pale as a sheet, and with livid lips. He groaned with pain, which he described as "cutting him across" from the sternal notch to the nipple, and going down the left arm; and there was so marked a catching of the breath as to make it almost certain that there was diaphragmatic spasm; in fact, it was this which alarmed him, and made him say that he was certainly dying. The heart, however, appeared to be pushed up somewhat, and it was thought that this might be partly due to stomachic distention, but a mustard emetic produced little effect. The heart-sounds were so weak that the presence or absence of bruit could not be safely predicated; meantime, the pulsations intermitted in a most alarming manner. Large doses of brandy and sulphuric ether at length (after several relapses) seemed to subdue the pain and spasm, and in an hour and a half from the commencement of the attack the patient, though utterly worn out, sank into a tolerably quiet sleep. The spasms did not recur, but for the next three or four days he was in a state of great exhaustion. When his tranquillity of mind had been somewhat restored, a careful physical examination was made, and it was discovered that there was a moderately loud and somewhat thrilling systolic bruit at the site of the aortic valves, and extending some distance into the vessels. The pulse still remained strikingly intermittent, and, though of fair volume, was very compressible. Percussion indicated considerable enlargement of the heart, and the physical signs pointed, on the whole, to dilatation without hypertrophy. Some doubtful signs of consolidation were observed at both apices of the lungs.

It is remarkable that, notwithstanding the serious degree of cardiac mischief indicated by the above signs, the patient, a very few days later, took a walk of some ten miles, and, though much exhausted, suffered no recurrence of his formidable spasmodic symptoms in consequence of this imprudence. He was sent to the mild climate of Mentone, and subsequently to Nice; the angina never recurred, but the patient remained weak, and liable to more or less dyspnœa for fifteen or sixteen months; now he lives an ordinary life, doing his duty as a Swiss citizen and officer. The cure of some hæmorrhoids, about twelve months after the anginal attack, seemed greatly to benefit him. What the future of this case may be it is impossible to say, but of course there is no security against the angina recurring on extraordinary excitement or overexertion.

Of the purely neurotic variety of angina it is impossible to determine the frequency; but it seems certain that the affection is common, and I suspect that it occurs more often than is supposed, as a sequel to asthma. The probable relationship between the two affections was long ago indicated by Kneeland.* I have certainly seen several cases of asthma in which spasmodic pain of the heart has occurred on various occasions after or during a very severe asthmatic paroxysm. One case was that of a gentleman, of a highly delicate and neurotic temperament, who had suffered for fifteen or sixteen years from well-marked spasmodic asthma: this case is remarkable as an illustration of several points which will be dwelt upon in other parts of this volume. For some time before the outbreak of cardiac neuralgia, he had suffered repeatedly from severe facial neuralgia, and these attacks on more than one occasion culminated in facial erysipelas, or what was entirely indistinguishable from that affection. He then began to suffer from cardiac pain and spasm after his asthmatic paroxysms, and these new symptoms speedily assumed the form of a very severe intermittent angina: in several of the attacks he appeared about to die. The pain in these attacks is very severe; it occupies a large area in the centre of the chest, and runs down both arms; and, what is strange, the arms become remarkably swollen and hot after an unusually long bout of pain, I presume from vaso-motor paralysis. At present (nearly five years from the commencement of the cardiac neuralgia) the cardiac attacks, though of frequent occurrence, are decidedly more tolerable than they were at first, and the sense of squeezing or pressure, though never quite absent, does not amount to the dreadful sort of feeling which used to convince the patient that he was at the point of death. In this case, the heart has been repeatedly explored without any positive result, and the pulse has been frequently tested by the sphygmograph. The latter instrument is the only mode of examining by which I have been able to elicit even suspicious evidence that there is any organic change of the heart; by means of it I have lately obtained some grounds for suspecting that there is slight dilatation of the heart, but it is uncertain whether anything of the kind existed at the commencement of the anginal symptoms. In this case I am inclined, on the whole, to doubt whether the angina will ever prove fatal, unless the bronchitis, with which the patient's asthma has for some time past been liable to be complicated, should occur in a severe form; in that case it is likely that the additional embarrassment of the heart's action may bring on fatal spasms.

One of the best examples I ever saw of cardiac neuralgia (ultimately proving fatal) was one of which the origin was en-

---

*Amer. Jour. Med. Science. Jan. 1850.

tirely nervous. It occurred in a gentleman in the prime of life, and naturally of a powerful physique, whose very active and capacious mind had been greatly overwrought. The whole weight of responsibility for an undertaking of national importance, and which involved great difficulties and much anxiety, for a long time rested on his shoulders. Under these influences he broke down, and never effectually recovered himself. At first, the symptoms were those of mere ordinary nervous exhaustion, but after a time he became subject to frequently recurring attacks of agonizing spasmodic heart-pain, with a sense of impending dissolution; from these he was invariably relieved by the inhalation of a small amount of chloroform. Not the slightest organic heart mischief could be detected, either during life or after death.

Pathology.—Angina stands in so peculiar a position that I deem it well to discuss it as a whole, and not merely its clinical history, in this place. As I have already said, there is nothing in the morbid appearances found after death which is characteristic of fatal angina, and in the milder kinds of cardiac neuralgia we are driven back upon the general probabilities which we deal with in reasoning as to the origin of neuralgias in general. As to morbid changes, it is impossible to say any thing more exhaustive of the facts known than the following words of Dr. Walshe :* "First, there are few, if any, structural diseases either of the heart, its orifices, and its nutrient arteries, or of the aorta, found recorded in the narratives of the post-mortem examination of different victims of angina pectoris. Secondly, there is no conceivable disease of these structures and parts which has not in various individuals reached the highest point of development, without anginal paroxysms, even of a slight kind, having occurred during life; to this proposition extensive calcification of the coronary arteries perhaps furnishes a solitary exception. Thirdly, the organic changes most frequently met with have been fatty atrophy and flabby dilatation of the heart; obstructive disease of the coronary arteries by atheroma and calcification of the orifice and arch of the aorta. Fourthly, the rarest have been hypertrophy and hypertrophy with dilatation. In truth, it may be doubted whether these conditions in their genuine form, without any combination of fatty atrophy, have ever been the sole morbid states present." From all this Dr. Walshe concludes that the fundamental mischief of angina is neurotic; and, while he believes that some textural change in the heart is necessary as an irritant to generate this neurotic susceptibility to dynamic disturbance from slight causes, he recognizes only one common quality in these various cardiac lesions, viz., that they indicate mal-nutrition and weakened

---

* "Diseases of the Heart and Great Vessels." Third edition, 1862.

power. Dr. Walshe does not appear to believe the neurotic disturbance can arise without the kind of irritation which is kept up by such cardiac changes. In spite, however of the great authority of this author, it certainly seems very probable that organic cardiac change is by no means necessary to the occurrence of angina, and this for two reasons: In the first place, though full reliance may be placed on the details of the post-mortem examinations made by Dr. Walshe himself, they are very few (twelve or fourteen) in number; and other observers who have recorded cases are as little trustworthy, considering their evident tendency to find some disease where none exists, as the older narratives which Dr. Walshe naturally distrusts were unreliable when they declared that no morbid change was present. And, secondly, his view hardly takes it into account that there are still two other alternatives, even supposing that one or other of the above changes is always present: (a) it is possible that the neurotic disturbance and the cardiac lesions might both be the result of a common cause; and (b) it is even possible that the alterations of tissue in the heart and vessels are due to a morbid influence proceeding from a diseased nervous centre, either spinal or sympathetic.

As for the state of the muscular fibre which immediately causes death, Dr. Walshe is of opinion that it is paralytic rather than spasmodic; and he urges in favor of this view the fact that in his large experience he has never known the pulse to intermit during the attack—it was always regular, however feeble. In this respect he is in opposition to some distinguished authors, however, and, as he allows that he has not seen original attacks in their height, but only when they were subsiding, it would be possible that the spasm stage had subsided. However Dr. Walshe admits that there may be exceptional cases in which spasm, or cramp (*i. e.*, spasm with rupture or dislocation of fibre), really occurs, and suggests that this is very probable in the rare cases where death is attended by general tetanic spasm of the muscles. As far as my own opinion is worth anything, I could insist that at least Dr. Walshe must be right as against Dr. Latham and Dr. Inman, in affirming that cardiac cramp, if it occurs, is the consequence and not the cause of the neuralgic pain.

Causes.—In some respects it is impossible to deal with the etiology of angina apart from the pathology, just as we remarked with regard to neuralgias in general. But there are certain special features in the causation of angina pectoris which require separate notice, just as there are special features in its pathology.

Of predisposing causes, the majority are the same as those of which we have spoken in our general remarks on the etiology of neuralgia. A family history of a tendency to the graver neuroses is I believe universal, and, indeed, direct inheritance of angina from father to son, as in Arnold's case, has happened in many recorded instances. A very remarkable fact is the time of life at which the disease originally appears: Walshe says it is rare before the age of fifty, but excessively rare before forty. This is very interesting, as placing angina in the same category with the severe and intractable forms of facial and other neuralgias which are so highly characteristic of the period of bodily degeneration. One may even gather a suspicion, though it goes but a short way toward proof, that the essence of angina is an atrophy either of the cardiac plexus or of the nucleus of the vagus, or of that part of the spinal cord, already mentioned, which seems to be the centre of the major part of the propulsive force of the heart.

On the other hand, there is a fact, even more remarkable than the influence of age, which tells somewhat in a contrary direction. There is a most extraordinary preponderance of males among the victims of angina. Sir John Forbes found eighty males among eighty-eight patients suffering from this disease. On the first blush it would seem natural, indeed almost necessary, to explain this by supposing that, as men take a much larger amount of strong physical exercise than women, they will furnish a much larger proportion of subjects in whom an ill-nourished heart will break down under its work and be seized either with paralysis or cramp (for the two states are, after all, not opposed to each other, but only varying shades of debility.) Upon this theory one would have to believe that the origin of angina was far more peripheral than central, if we are to suppose that spasm is the ordinary condition of the heart during the anginal paroxysm. But we do not know that this is the case; indeed, there are many arguments against it; and at any rate we must suppose that in a considerable number of cases the muscular state is one of relaxation from want of power. And certainly it is infinitely more probable that paralysis or spasm of a muscular viscus should occur as a reflex consequence of neuralgia occurring in a nerve whose central nucleus was closely connected with the motor centre of the organ, than that mere paralysis of the viscus should convey a reflex impression to sensitive nerves which should express itself in the form of acute pain. It must be confessed that the matter hangs in doubt; but the evidence is, on the whole, very strong for the belief that central nervous mischief is the most important element in angina.

Another very important class of predisposing causes of angina is the mental emotions. It is notorious that the disease is one not common in humble life; it chiefly assails the more cultivated class, and especially men who are much engaged in affairs in which great mental anxiety or emotion is mingled with severe toil of intellect. Thus the professional class has always shown a sad predominence in tendency to this disease;

a large number of the victims have been found among overworked clergymen, lawyers, doctors, engineers, etc. The various forms of heart-lesion which have been already mentioned must doubtless be considered highly predisposing, when there is already a neurotic susceptibility, more especially those which, like fatty degeneration of the muscular structure, greatly enfeeble the heart's action. I do not believe that these diseases will cause angina in a person who is free from the peculiar nervous susceptibility.

The immediately exciting causes are very various. The most common of all is doubtless some exertion of body, or distress of mind, which at once agitates and embarrasses the heart's action; and, where the tendency to cardiac neuralgia has once declared itself by an actual attack, very slight excesses of this kind will usually suffice to re-excite the paroxysm. Sexual excitement is particularly provocative of the attacks, in the predisposed. But much slighter causes suffice, in those cases where the irritability of the cardiac nerves has become very intense: thus a mere puff of cold air upon the face, and other similar slight peripheral impressions, by acting in a reflex manner, have frequently produced the paroxysm. I have seen an extremely severe anginal attack brought on by the slight shock of the sudden slamming of a door. And it would even appear that some peripheral excitements of a powerful kind may operate with such force as to generate angina in persons who are merely in weak health, but who cannot be supposed to be specially predisposed to angina; it is in this way, I presume, that we must explain the extraordinary occurrence, reported by Guelineau,* of an epidemic outbreak of angina, in which numbers of men, belonging to a ship's crew, were simultaneously affected. The men had been badly fed, and their quarters were very unhealthy; but the powerful exciting cause seemed to be the rapid change from a very hot to a very cold climate. Not only were there many cases of severe angina, but other forms of neuralgia, and severe colics, were observed in others of the crew. Among the sources of peripheral irritation which ought to be particularly considered, in relation to angina, are the diseases and injuries which produce powerful irritation of the branches of the trigeminus. Lederer's cases † of violent vomiting and cardiac pain, from the operation of pivoting teeth, and Remak's instances ‡ of violent palpitation and cardiac distress, produced by disease of the last molar tooth, seem to show that, both through the vagus and the sym

---

*Gaz. des Hop., 114, 117, 120. 1862.
† Wien Med. Presse, xxiv., 1866; Syd. Soc. Yearbook, 1865-'66, p. 120.
‡ Berlin Klin. Woch., 1865; Syd. Soc. Yearbook, 1865-'66, p. 120.

pathetic, the most powerful reflex action may be produced in the heart and stomach by irritation of the fifth cranial.

Another occasional excitant of angina is an interesting link in the chain of proof that angina is *au fond* a neuralgia, namely, the malarial poison, which has in a good many well-observed cases distinctly induced the disease.* Finally, the occasional influence of excessive tobacco-smoking in producing anginal attacks, in persons not affected with any discoverable organic heart-disease, affords the strongest corroborative evidence of the essentially neurotic character of angina pectoris. M. Beau† has recorded many serious, and some fatal, cases from this cause. Probably in both the malarial cases and those induced by tobacco-poisoning the special neurotic tendency existed already.

Diagnosis.—The diagnosis of angina pectoris, in those severe forms with which the popular idea of the disease is chiefly connected, can hardly be a matter of much difficulty. When we see an elderly man lying in a state of deathly collapse, which has suddenly come on, with cold sweats and nearly extinguished pulse, gasping for breath, and complaining of intolerable pain in the chest and arm, and a sense of oppression more dreadful, even, than the pain, we can hardly doubt that the case is angina in its worst form. On the other hand, when a young person, especially a young female, complains even of very severe pain in the cardiac region, together with breathlessness, especially if the heart be palpitating and the face flushed, the diagnosis, though not immediately certain, already very strongly indicates the probability that the case is not one of primary cardiac neuralgia at all. These are extreme instances, however. In more doubtful cases, the following are the principal materials for decision:

Affirmative Signs.	Negative Signs.
1. Age over forty.	1. Age under forty.
2. Male sex.	2. Female sex.
3. Nervous temperament (personal and family) without marked hysteria or hypochondriasis.	3. Temperment either not nervous at all, or markedly hysterical or hypochondriacal.
4. Existence of arterial degeneration.	4. No signs of arterial degeneration.
5. Existence of valvular disease of the heart.	5. No discernible valvular disease.
6. Extension of the pain to one or both arms.	6. Heart sounds clear and strong.

---

* See Wahn, *Journ. de Med. et Chir. Prat.* 1854. Also several original and quoted cases in Dr. Handfield Jones's "Functional Nervous Disorders," second edition, 1870.

† *Journ. de Med. et Chim. Prat.*, July, 1862.

Affirmative Signs.	Negative Signs.
7. Vivid sense of approaching dissolution.	7. Pain fixed to one spot and increased or relieved by muscular movements of the painful parts.
	8. Pain running round one side, but not extending to shoulder or arm.

It is scarcely necessary to say that no single one of the above signs is individually of positive worth for the decision, which must be made after a careful review of the comparative arguments, *pro* and *con*. The disorders with which angina is most likely to be confused are (1) Myalgia of the intercostal or pectoral muscles; (2) intercostal neuralgia; (3) acute commencing pleurisy. Either of these may very perfectly simulate the more formidable disease, as regards the two elements of acute pain and catching of the breath; but the condition of the circulation, taken together with the consideration of the above named points, will generally decide the question. Especially important is the deep persuasion of impending dissolution, when present, as a positively affirmative symptom.

It should be born in mind that, if we are summoned to a patient's assistance, and have no previous history to guide us, our diagnosis, to be useful, must be rapid; and it is always better to err on the side of angina than in other directions, and to employ remedies boldly in that sense, if there be any reasonable ground for believing the case to be of that nature. A more mature and careful diagnosis may be made when the patient has recovered from the severe symptoms of the paroxysm.

Prognosis.—The prognosis of cardiac neuralgia is at best doubtful, and, in many cases, positively bad in the highest degree. If the attacks occur for the first time in a patient who has passed middle life, and is physiologically old for his age, *i.e.*, shows tendency to degenerative changes of vessels, arcus senilis, gray hair etc., they are of very gloomy import; more especially if any signs exist which make a fatty change in the ventricle probable, or if there be serious valvular lesions. The probability here is greatly in favor of a speedy fatal termination; if the first attack does not kill, a second or third very probably will; at any rate, the patient is not likely to survive any considerable number. If the attack occurs in a younger person, in whom there is not much likelihood that arterial degeneration has seriously commenced, or the heart-muscles become fatty, more especially if the attacks have been brought on by such an accidental circumstance as a very exhausting bout of mental or physical toil, then there is considerable reason to hope that the disease may soon wear itself out. Even patients who have serious valvular lesions may, with young and undegenerated tissues in their favor, quiet down again into a regular habit of semi-health, in which they may live for a long time without any recurrence of cardiac neuralgia. The more purely neurotic form, again, especially when it develops gradually out of some pre-existing chronic neurosis, such as asthma, is usually slow in its progress; and it may well happen, in such cases, that the danger to life is more on the side of serious nervous lesions than from the anginal attacks themselves. At the same time, it must be remembered that, even in the milder cases, any very unusual excitement, bringing on an unwontedly severe attack, may produce fatal results at any period of the disease.

There is some reason to believe that cardiac neuralgia is occasionally produced in a reflex manner in consequence of a severe existing intercostal neuralgia. I cannot say that I have witnessed any thing which can be considered as completely proving this; but it certainly seems likely that, in some of the few cases of excessively painful herpes zoster which have proved fatal (of which I have given one example), cardiac spasm or paralysis may have been secondarily induced, and may have occasioned the catastrophe. It is likely enough that, if this was the case, the reflex irritation operated upon motor centres which themselves were predisposed to take on the morbid action; but this again is a fresh illustration of the uncertainties to which prognosis is liable in a disease like angina, the very fundamental character of which is that, upon increase of the irritation, the gravity of the resulting functional affection is liable to be indefinitely and most rapidly increased.

Treatment.—The treatment of cardiac neuralgia is (1) prophylactic, and (2) palliative of the attacks.

As regards the prophylactic treatment, it is unnecessary to repeat the remarks which we have made elsewhere upon the general principles of tonic and nutritive medication in neuralgias of every kind. One especial prophylaxis, in the case of this formidable variety of neuralgia, is concerned with the preservation of the heart from certain disturbing influences which would render the occurrence of the fit more probable. All violent emotions and all strong physical exercise (but especially such forms of it as, like boating, are well known to "pump" the heart severely) are to be carefully avoided. Even indigestion and flatulence are to be carefully guarded against since these are quite capable of embarrassing the action of the heart to a degree which, though it might be trivial in the case of ordinary health, may prove fatal by exciting a flabby ventricle to irregular and embarrassing contraction. It is even possible that the strong irritation set up by some varieties of indigestible food might propagate an irritation to the spinal cord which would produce an interbitory paralysis at once.

But besides these obvious precautions against interference with the regular and tranquil action of the heart, there are some special medicinal remedies which deserve particular notice. Whether we really possess any means of so influencing the nutrition of the muscular tissue of the heart as to prevent its lapsing into a fatty degeneration, it is impossible to say; but this may be affirmed with some confidence, that, in cases where awkward threatenings of cardiac neuralgia have occurred, and simultaneously it has been noticed that the heart-sounds become weak and the circulation languid, a most marked improvement has been produced in all respects by the administration of iron and strychnia. I usually give tincture of sesquichloride of iron, ten minims, and strychnia, one-fortieth of a grain, three times a day. Still better, where it can be borne, is the syrup of the triple phosphate of quinine, iron, and strychnia, which undoubtedly has an extraordinary influence upon tissue nutrition, as exemplified in its remarkable effects in many cases of phthisis. It must be observed, however, that it is not every neuralgic patient who will bear the combination of quinine with iron; it has occurred to me to meet with several in whom the union of these two remedies proved violently disturbing to the nervous system, causing distressing headache and palpitation of the heart, which could not be attributed to any want of care in the apportioning of the dose, or in the mode of administration. Iron is more especially indicated, of course, in cases where there is anæmia; but there are some cases in which strychnia given alone seems to produce a very beneficial influence. (*Vide* Chapter V., on "Treatment.")

By far the most important prophylactic tonic against cardiac neuralgia, however, is arsenic. That this drug should prove useful in cardiac neuroses might readily be anticipated from its very great utility in many cases of asthma, a disease which, as already remarked, has a close relationship to the former. Dr. Philipp has recently recorded a case which is perhaps an extreme instance of this beneficial influence of arsenic, but is none the less encouraging, especially as it only corroborates what has been advanced by other observers. Given in doses of from three to five minims of Fowler's solution, twice or thrice daily, arsenic is an invaluable remedy in cardiac neuralgia ; the one objection to it being that some neurotic patients possess such an irritable intestinal canal that the remedy cannot be borne, as it produces diarrhœa. Even here we may sometimes succeed by combining it with very small doses of opium. It is more especially with regard to those cases in which the neurotic character of the disease is very prominent—*i.e.*, in which the nervous temperament of the patient betrays itself in other ways besides the tendency to spasmodic embarrassment of the heart's action, that arsenic holds such a very high place as a remedy. And it should be carefully remarked that the prophylaxis of angina extends itself, in such cases, beyond the limits of actually-declared and well-defined angina, which is, of course, an uncommon disease. This remedy is important, and may be most usefully employed in the far larger group of cases in which a marked tendency to spasmodic pain in the chest, on the occurrence of some comparatively trifling excitement, is observed in patients who either have some organic heart-disease, or who are liable to severe attacks of asthma. It cannot be too often repeated that there is no intelligible separation, except one of degree, between these cases and the malignant forms of angina. It may be added that, in my experience, I have found the whole group of cases to be bound together in a singular way by the tolerance of arsenic which, with certain exceptions already referred to, they display. Commencing with the small doses above mentioned, I have found it possible, in many cases, to advance to the administration of twice or thrice the quantity, and to continue this medication for months together, not only with no evil effect, but with the best results.

Of zinc, as a prophylactic tonic in cardiac neuralgia, I know but little. Truth to say, it is a nervine tonic of occasional great value, but which, on the whole, I have found so unreliable that I am somewhat prejudiced against it; and perhaps have not given it a fair trial in those milder cases of cardiac pain to which it might be suited. It does appear, however, to have some preferential action on the vagus, and might therefore be possibly more useful than I am at present inclined to think it.

The treatment of the acute neuralgic stage itself is a matter in which we are sadly limited by the exigencies of the case. Relief must be excessively rapid if we are to save life in the most threatening cases, or to deliver the patient from a most prostrating agony, which might have lasted for hours, in other instances.

The remedy which the highest authority, Dr. Walshe, seems to put first in efficacy is opium; and he directs the dose to be measured by the intensity of the pain, as much as forty to sixty drops of laudanum being given in a severe case. He says, however, that it should be given with an antispasmodic, such as brandy, or ether, or sal-volatile ; and I confess that I believe the antispasmodic treatment to be by far the most important. Indeed, so marked is the success which I have found to attend the use of ether in the paroxysm, that till lately I scarcely cared to make further experiments, with drugs, for the relief of the patient at this stage. One teaspoonful of ether in two ounces of thickish mucilage should be given at once, and repeated in a short time if the patient does not rally.

In a few instances, angina seems to be provoked by the irritation of indigestible food, and when there is good reason to suspect this an emetic should be given. I strongly recommend that mustard should be used for this purpose, for the effect of a mustard-emetic is by no means merely to empty the stomach, it has a powerfully rousing influence on the heart.

Upon the subject of the inhalation of chloroform for cardiac neuralgia, I have only to say that, though I have seen it usefully employed, I should not, with my present experience, ever think of employing it myself. Every possible advantage which it could give is obtained by the internal use of ether, and many serious dangers are avoided, which would attend the use of chloroform. For it must be remembered that the only kind of chloroform inhalation which would be useful would be that in which a carefully measured small dose of a weakly impregnated atmosphere should be inhaled, and, without large experience in the administration of chloroform, the practitioner will be unable to secure this effect with certainty. And the effect of a powerfully-charged atmosphere, breathed only once or twice even, would be instantaneously fatal.

Hot epithems to the epigastrium are probably of some use, and besides this the temperature of the body should be carefully kept up by hot bottles to the feet, hot tins to the epigastrium, etc. Brandy should be freely administered during the attack, if we cannot immediately obtain either ether or a remedy now to be mentioned. I refer to the nitrite of amyl, which, at the time when the first part of this chapter was written, I had not had the opportunity of testing.

Nitrite of amyl is a highly-vaporizable fluid, which possesses the following remarkable physiological action : the inhalation even of a very small quantity is followed, after a minute or so, by a sudden acceleration of the heart's action, accompanied by intense crimson congestion of the vessels of the face and conjunctiva, and a sense of enormous fulness in the head; these phenomena are extremely fugitive, passing away completely in two or three minutes, unless the inhalation is renewed. These characteristic effects had for some years been experimentally exhibited by Dr. Fraser and others, but the practical application of amyl to the treatment of angina was first suggested, I believe, by Dr. Brunton, in the case of a patient under the treatment of Dr. Maclagon and Dr. Bennett, in the Edinburgh Royal Infirmary. The angina was in this case symptomatic, there being advanced valvular disease of the heart. Comparative examinations with the sphygmograph, during the intervals and during the paroxysms, made strikingly manifest the fact that, during the attacks, there was an increase of arterial tension which was directly proportionate to the severity of the pain and cardiac embarrassment. It was thus suggested to Dr. Brunton's mind that nitrite of amyl, by relaxing the systemic arteries, might remove the unnatural tension, and give relief to the pain; and the result confirmed this hope. Doses of five and ten drops were inhaled from a towel, with the uniform result of at once quieting the pain; it might return in a few minutes, but a second dose usually removed it entirely for many hours. Various other cases have since been reported, in which similar relief was obtained, and I had occasion to employ it myself in one instance. The gentleman whose case has been related above (see page 101), as an example of the relief obtainable by the use of ether began to suffer rather more severely from his attacks than had been the case for some time, toward the end of the year 1869. I now determined to try the amyl, and accordingly left a small bottle containing half an ounce of it in his possession, with exact instructions to the following effect: On the first symptoms of a paroxysm of angina, he was to get the bottle open, and as soon as their character was fully declared he was to put the bottle to one nostril (closing the other with the finger, and keeping the mouth shut) and take one long, powerful inspiration. The result of his first experiment was very remarkable: the first sniff produced, after an interval of a few seconds, the characteristic flushing of the face and sense of fulness of the head; the heart gave one strong beat, and then at once he passed from the state of agony to one of perfect repose and peace, and at his usual bedtime slept naturally. This experience was repeated on several occasions, and for a considerable time the patient retained such full confidence in the remedy that he discarded all use of ether, and greatly reduced his allowance of stimulants, with very marked benefit to his appetite and general health. The new remedy did not lose any of its power by repetition, but unfortunately the patient at last conceived a horror of it, which caused him to abandon its use. So distressing and alarming to him was the sense of fulness in the head produced by the amyl, that, notwithstanding his certain knowledge that he could at once cut short a paroxysm, he could not persuade himself to continue its use, and for some time past he has returned to the use of the ether and (though in less quantities than previously) of the brandy, for this purpose. And here it must be remarked that this objection, although probably needless in the case of this particular patient, may have real importance in certain circumstances. The admirable physiological researches of Dr. Brunton leave no doubt that the effect of inhalation of amyl is to relax, very suddenly, the tonic contraction of the systemic arteries, and in the case of the brain it would appear that a serious strain must be suddenly thrown upon the capillary net-work. This being the case, it appears likely that, where the atheromatous change has considerably invaded these delicate vessels, they might prove too brittle to stand the sudden distention, and a rupture and consequent cerebral hæmorrhage might ensue. This suspicion, then, that such pathological changes exist, ought to seriously affect our judgment as to the administration of amyl; and this suspicion ought to be always entertained, *prima facie*, in the case of patients who have much passed the age of fifty, more especially if they have gray hair and an arcus senilis, or if the sphygmograph yields a pulse-trace of the decidedly square-headed type, or if they have been long addicted to alcoholic intemperance. In such patients I should be disinclined to allow the use of amyl.

[Although I have thought fit here to give an outline of angina pectoris as a connected whole, I shall have occasion to recur to the subject again under the heads of Pathology and Treatment of Neuralgias in General.]

Gastralgia.—Neuralgia seated in the stomach itself is not to be distinguished with accuracy from neuralgic pains occupying one or other of the neighboring nervous plexuses. It must be remembered that not merely is the stomach itself copiously supplied by the pneumogastric nerves with afferent fibres, but the great solar plexus is close behind it, the cœliac plexus springs from the fore part of the latter, and these, with the coronary and superior mesenteric plexus, may all be said to be well within the region in which "gastralgic" pain is felt. It is not particularly important, however, in my opinion, to make any very exact diagnosis here, as to the site of the pain, since all these neuralgias must be considered to belong to the pneumogastric nerve, the branches supplied from which are probably the sole means by which these plexuses become the seat of neuralgia.

Abdominal pneumogastric neuralgia is an extremely distressing and occasionally a very intractable disorder. The subjects of it are almost invariably in a state of marked and evident debility, and inquiry generally elicits the fact that they have suffered at other times from neuralgia elsewhere than in its present seat. By far the most common history of previous affections of this kind is that of trigeminal neuralgia, especially of the supra-orbital branch; and it has several times occurred to me to observe the direct sequence of a gastralgia upon a unilateral browache. Anæmia is a specially frequent attendant of gastralgia, more so than of other neuralgias. Women are, by the general consent of authors, more liable to gastralgia than men.

The special mark of true neuralgic pain in the abdominal pneumogastric, as distinguished from other deep-seated pains in the epigastrium, is the remarkably direct relation of its severity to the patient's exhaustion, particularly in regard to the weakness induced by want of food. While the great majority of dyspeptic pains are increased by filling the stomach, gastralgia, on the contrary, is invariably relieved by food, often most strikingly and completely. Pressure from without, also, while it aggravates most pains dependent on local organic mischief, nearly always more or less relieves gastralgia. Equally striking is the comfort given by stimulants, especially by hot brandy-and-water; in this respect gastralgia resembles colic. There is something special in the degree of mental depression which attends gastralgic pain. In this it resembles the pains of hypochondriasis, but there is a resilience of the spirits when the pain has been relieved which is not seen in the latter affection. A very frequent complication of gastralgia is severe palpitation of the heart, but during the paroxysm itself the pulse, whether rapid or not, is commonly small, at first tense, and afterward soft, but not acquiring any considerable volume till the pain has ceased.

So severe is the pain, and so complete the mental and physical prostration in bad attacks of gastralgia, that the first aspect of the patient might suggest—indeed often has suggested—the occurrence of gastric or duodenal perforation; but, as soon as the paroxysm is over all the alarming appearances vanish, leaving only a certain amount of tenderness on deep pressure. In the more typical cases there are no signs of dyspepsia whatever, no fulness nor excessive redness of the tongue, no nausea, regurgitation of food, nor pyrosis. Occasionally the neuralgic affection is complicated with more or less gastric catarrh; but this is a much rarer occurrence, in my experience, than some writers would lead one to believe; and, moreover, where a certain amount of organic disorder of the stomach is observed, it is usually a mere secondary result of the neuralgia. The most severe example of gastralgia which I ever saw was entirely unaccompanied by dyspepsia; this patient absolutely attempted suicide to escape from his agonizing pains, which recurred with the greatest frequency and obstinacy, but were at last entirely removed by strychnia, In another patient whose very interesting case will be again alluded to under the head of Complications of Neuralgia, violent abdominal pneumogastric pain was succeeded by a severe attack of trigeminal neuralgia, accompanied by inflammation of the eye, which inflicted irreparable damage; here, too, the gastralgia was entirely uncomplicated by any other stomach-symptoms.

Cerebral Neuralgia.—We enter, here, on an extremely obscure and doubtful subject: Can there be pain in the central masses of the encephalon? There are undoubtedly a not inconsiderable number of cases of pain, neuralgic in type on the whole, in which the suffering cannot be referred to any recognizable superficial nerve. It seems deeply situated within the cranium. I have also quoted cases of Dr. Hillier's in which not merely was there deep-seated headache in children, but there was something like a characteristic general change observed in the brain-tissues after death, viz., a great moisture and softness of texture. Notwithstanding all this, I am not convinced, nor indeed much disposed to believe, that pain is ever felt in the structure of the brain; I rather believe that, in the cases where this seems to occur, the pain is either in the intracranial portion of the nerve trunks, or, far more probably, in the twigs of nerves that are distributed to the cerebral membranes. In that case they are, strictly speaking, only varieties of neuralgia of the fifth nerve, and might have been properly discussed under that heading; but it is more convenient to speak of them apart, since their phenomena present considerable differences from those of the external neuralgias of the head and face.

I have now seen several of these cases of intracranial neuralgias, and very perplexing and (at first sight) alarming they certainly are. The first of these cases came under my care in 1868. The patient was a single lady who had greatly overtasked an intellect that was not, perhaps, originally very strong, by trying to do hack literature on conscientious principles; insisting, for instance, on knowing something about every subject she wrote upon. Her age was thirty-eight when she applied to me; menstruation was scanty but regular; and, on the whole, she could not be said to have passed an unhealthy life, although "nervous-headaches" and "sick-headaches" had occasionally beset her. This time the trouble seemed to be more serious. Ten days before applying to me, she had awaked in the morning with a feeling that something was very wrong in her head; there was not so much pain as a dull, brooding sort of weight, felt deeply within the cranium, and rather anteriorly. This had not lasted many hours when she was seized with a sensation of intense cold, amounting almost to rigors, and then before long was suddenly attacked with acute splitting pain in the same situation as the feeling of weight already mentioned had occupied. This pain, which came and went, or rather intensified and remitted, without ever completely ceasing, lasted about two hours, and then rather suddenly disappeared, leaving the patient with a deep "bruised and sore feeling in her brains." The pain recurred about the middle of the next day, lasting for several hours, and again leaving behind it the sore feeling. Day by day the paroxysms returned, and, on the day before her visit to me, the patient had, she told me, been driven frantic by her sufferings and had become actually delirious. Her appearance, when I first saw her, was wretched; the face haggard, both eyes sunken and surrounded with deep rings of dusky pigment, both conjunctivæ bloodshot, the whole face almost earthy in its pallor. At that hour (11 A. M.) the pain had not positively recommenced, but she was in momentary dread of its recurrence. She complained of giddiness, muscæ volitantes, and great feebleness of vision, and dreaded attempting to read, as the mere effort of fixing her eyes on anything intently caused flashes of fire before them. It was difficult at first to believe that there was not some serious organic brain-mischief; but on the whole I concluded that there was an absence of any genuine symptoms of such disease. At the same time, the pain was decidedly not referred to any cutaneous sensory nerve; and on the whole it appeared probable that the affection was intracranial. There remained the diagnosis of meningeal neuralgia, and to this I provisionally made up my mind. The opinion that the pain did not depend an any fixed organic disease was decisively justified by the results of treatment. One-sixth of a grain of morphia was injected on the occasion of the first visit, and this was repeated every day, and sometimes twice a day, for a fortnight; by this sole means, with rest, quietude, and light nourishing food, the patient was brought to comparative convalescence. The injections were then gradually discontinued, and she got quite well.

In a second case, which presented itself in the out-patient room at Westminster Hospital, a young man of markedly-nervous temperament, who had been somewhat given to drink, complained of similarly deep-seated intermittent pain, which he referred, however, to a point nearer the back of the head. He suffered, also, from vertigo, especially after unusually long paroxysms. Blisters to the nape of the neck, and a few subcutaneous injections of morphia, removed the pain and the vertigo completely.

A third example was that of a gentleman, aged thirty-four, who was sent over from the neighborhood of Sydney, Australia, to see me. Here, also, there was deep-seated intracranial neuralgic pain of the most severe kind, which greatly alarmed his local medical attendants; and it was only after a great many remedies had been tried that one medical man gave the opinion that the disease was "neuralgia of the membranes of the brain," and employed the hypodermic injection of morphia. This treatment at once gave great relief, though the pain had been so severe as to cause delirium on several occasions. In order to get thoroughly re-established, he was sent to England, and desired to consult me. As was expected, the voyage proved of the greatest service, as he hardly suffered at all while on the water. On arriving in England he was at first well, but in a week or two began to feel somewhat below par, and one morning, feeling an attack of pain coming on, he came to me. He was a tall and strongly-built man, with nothing peculiar in his appearance except a certain languor and heaviness of the eyes. He appeared to have lived somewhat freely and to have smoked decidedly to excess. His description of the attacks left no doubt of their neuralgic character, and in other respects they seemed quite analogous to the other cases mentioned above, except in one thing, that there seemed a good deal of evidence tending to show a bad local influence in the air of that part of Australia where he usually resided. Almost any change from that had always done him good, though nothing had done anything like so much as the voyage to England. On the occasion of his first visit to me I injected him with one-sixth grain acetate of morphia, thereby stopping the pain. I prescribed muriate of iron and minute doses of strychnia, which he took for some little time, but the pain never recurred during his stay in England and on the Continent. Unfortunately, as he was anxious to return to Australia, I permitted him to do so, after a stay in the Old World of only three or four months; but, very shortly indeed after his return to Sydney, his old complaint attacked him. This time, unhappily, the hypodermic morphia has proved merely palliative, and I have latterly heard very bad accounts from him; still, there has been nothing to throw doubt on the neuralgic character of the disease.

In reflecting upon the anatomy of the nervous branches to the dura mater, I have formed the opinion that there are two situations, one anterior and the other posterior, in which intracranial neuralgia may occur; the former at the giving off of Arnold's recurrent branch from the ophthalmic division, near the sella turcica, the other in the peripheral twigs of this same branch, distributed to the tentorium cerebelli.

Pharyngeal Neuralgia.—A rather common and extremely troublesome form of neuralgia is that which attacks the pharynx. It is very much more common in women than in men, and especially in hysterical persons. The pain commonly commences in a not very acute manner; it may be felt for some days, or even weeks, as a dull aching, coming and going pretty much in accordance with the patient's state of fatigue, or of reinvigoration after meals, etc. Some trivial circumstance, such as a slightly extra degree of exhaustion, or the influence of some depressing emotion, will then change the type to that of decided neuralgia, which may become extremely severe. Nothing is more annoying, and even distressing, than the suffering itself, besides which there are abnormal sensations in the throat which almost irresistibly compel the patient to believe that there are severe inflammation and ulceration, and that the throat is in danger of being closed up. Although the pain is usually one-sided, it sometimes affects both sides, and is felt also at the back of the pharynx. The act of swallowing being painful, there is the greater suspicion of inflammation or ulceration, but careful observation shows that a large bolus of food is swallowed with as little, if not less, pain than a small mouthful of solids or even liquids.

Pharyngeal neuralgia must, I think, be considered mainly an affection of the glosso-pharyngeal nerve; the evidence for this is found in the distribution of the pain. A slight degree of the neuralgia will only involve some one or two points in or behind the tonsil; but, when the pain is strongly developed, it will be found to radiate into the tongue, in one direction, and into the neck (following the course of the carotid) in another, besides spreading well into the region occupied by the pharyngeal plexus. One disagreeable reflex effect of severe pharyngeal neuralgia consists in involuntary movements of the muscles of deglutition, another is seen in the copious outpouring of thick mucus similar to that which collects in the pharynx and œsophagus when a foreign substance has become impacted.

Laryngeal neuralgia concentrates itself mainly in the twigs of the superior laryngeal branch of the pneumogastric which are distributed to the arytæno-epiglottidean folds, the epiglottis, and the chordæ vocales; more rarely a neuralgia is developed lower down, within the cavity of the larynx, apparently in one or more of the scanty twigs to the mucous membrane supplied by the recurrent laryngeal.

Pure neuralgias of the larynx, like those of the pharynx, are more common in women, and especially in weakly hysterical women, than in men. They are easily excited and greatly aggravated by movements of the parts, and thus it happens that, among men, by far the most numerous subjects of laryngeal neuralgia are found among clergymen, professional singers, and others whose occupation compels them to strenuous and fatiguing employment of the laryngeal muscles. It is rather a singular and striking fact, however, that the so-called "clergyman's sore-throat," which is characterized by most unpleasant sensations, and by a more or less complete loss of voice, is not, in the majority of cases, attended with any distinct laryngeal neuralgia. It seems that a predisposition to neuralgia is a necessary element in the latter affection.

CHAPTER II.

COMPLICATIONS OF NEURALGIA.

The secondary affections which may arise as complications of neuralgia form a deeply interesting chapter in nervous pathology, and one which has only been explored in quite recent years. The excellent treatises of Valleix and Romberg, written only thirty years ago, make but most cursory and superficial mention of these complications, and do not attempt to group them in a scientific manner. The reflex convulsive movement of the facial muscles in severe tic-douloureux had of course been long observed; and Valleix added the correct observation that gastric disturbance was often secondarily provoked in facial neuralgia, thus improving greatly on the old view, which supposed that, where trigeminal neuralgia and stomach disorder coexisted, the latter must have been the antecedent and the cause of the former. Still, he did not explain the pathological connection. And as regards certain other most interesting results of neuralgia, which he could not avoid meeting with from time to time, *e. g.*, lachrymation, flux from the nostril, salivation, altered nutrition of the hair, he only speaks of these as occasional phenomena, and in no way classifies them, or explains their relation to the neuralgia itself.

There did exist, however, one too little known work of some years earlier date, which, though not dealing specifically with neuralgia, and though based upon the necessarily very imperfect knowledge of the functions of the nervous system prevalent in its day, had nevertheless done much to lay the foundation of a comprehensive view of the complications of neuralgia; we refer to the work of the brothers Grffin, on "Functional Affections of the Spinal Cord and Ganglionic System," published in 1834. In this most interesting treatise, the record of acute and extensive observations made in a quiet and unpretending way by two Irish practitioners, numerous examples are cited in which neuralgic affections were seen to be inseparably united with secondary affections of the most various organs, with which the neuralgic nerves could have no connection except through the centres, by reflex action. The authors, while firmly grasping the fact of the common connection of the nerve-pain and the other phenomena (convulsions, paralysis, altered special sensation, changes in secretion, changes even in the nutrition of particular tissues) with the central nerve system, were doubtless in error in thinking that they could detect the precise seat of the original malady, by discovering certain points of tenderness over the spinal column. But their facts were observed with the greatest care, and can now be interpreted more intelligently than was possible at the time. Here, for example, is a case which forestalls one of the most interesting pieces of information which more recent research has made generally known:

"CASE XXIV.—Kitty Hanley, aged fourteen years, catamenia never appeared; about six months ago was attacked with pain in the right eye and brow, occurring only at night, and then so violently as to make her scream out and disturb every one in the house; it afterward occurred in the infraorbital nerve, and along the lower jaw in the teeth, and there was inflammation of the cornea, with superficial ulceration and slight muddiness. Tenderness was found at the upper cervical vertebræ, pressure on any of them exciting severe pain in the vertex and brow; but none in the eye or jaws, where it is never felt except at night."

The above is a well-marked example of neuralgia of the trigeminus causing secondary inflammation and ulceration of the eye of a precisely similar kind to that which had been experimentally produced by Magendie by section of the fifth, at or posterior to its Gasserian ganglion. We shall see, hereafter, how extremely important are this and similar facts, not only in regard to the clinical history, but also to the pathology of neuralgia in general.

The first regular attempt, I believe, to classify the complications of neuralgia, was made by M. Notta, in a series of elaborate papers in the "Archives Generales de Medecine" for 1854. We may specially mention his analysis of a hundred and twenty-eight cases of trigeminal neuralgia, which is well fitted to impress on the mind the frequency, though, as we shall presently see, it does not adequately represent the seriousness, of these secondary disorders. As regards special senses, Notta says that the retina was completely or almost completely paralyzed in ten cases, and in nine others vision was interfered with, partly, probably, from impaired function of the retina, but partly, also, from dilatation of the pupil or other functional derangement independent of the optic nerve. The sense of hearing was impaired in four cases. The sense of taste was preverted in one case, and abolished in another. As regards secretion, lachrymation was observed in sixty-one cases, or nearly half the total number. Nasal secretion was repressed in one case, in ten others it was increased on the affected side. Unilateral sweating is spoken of more doubtfully, but is said to have been probably present in a considerable number of cases. In eight instances there was decided unilateral redness of the face, and five times this was attended with noticeable tumefaction. In one case the unilateral tumefaction and redness persisted, and were, in fact, accompanied by a general hypertrophy of the tissues. Dilatation of the conjunctival vessels was observed in thirty-four cases. Nutrition was affected as follows: In four cases there was unilateral hypertrophy of the tissues; in two, the hair was hypertrophied at the ends, and in several others it was observed to fall out or to turn gray. The tongue was greatly tumefied in one case. Muscular contractions, on the affected side, were noted in fifty-two cases. Permanent tonic spasm, not due to photophobia, was observed in the eyelid in four cases, in the muscles of mastication four times, in the muscles of the external ear once. Paralysis affected the motor oculi, causing prolapse of the upper eyelid, in six cases; in half of these there was also outward squint. In two instances the facial muscles were paralyzed in a purely reflex manner. The pupil was dilated in three cases, and contracted in two others, without any impairment of sight; in three others it was dilated, with considerable diminution of the visual power. Finally, with regard to common sensibility, M. Notta reports three cases in which anæsthesia was observed. Hyperæsthesia of the surface only occurred in the latter stages of the disease.

To Notta's list of complications of trigeminal neuralgia must be added the following, all of which have been witnessed, and several of them in a large number of instances: Iritis, glaucoma, corneal clouding, and even ulceration; periostitis, unilateral furring of the tongue, herpes unilateralis, etc. In writing on this subject three or four years ago, I mentioned that all these secondary affections had been seen by myself, except glaucoma. That is now no longer an exception; indeed, my attention has been so forcibly called to the connection between glaucoma and facial neuralgia, that I shall presently examine it at some length.

The trigeminus is, of all nerves in the body, that one whose affections are likely to cause secondary disturbances of wide extent and various nature, owing to its large peripheral expanse, the complex nature of its functions, and its extensive and close connections with other nerves. Moreover, its relations to so important and noticeable an organ as the eye tends to call our attention strongly to the phenomena that attend its perturbations. But there is every reason to think that all secondary complications which may attend trigeminal neuralgia are represented by analogous secondary affections in neuralgias in all kinds of situations; and we may classify them in the principal groups which correspond to disturbance of large sets of functions:

1. First, and on the whole, probably, the most common of all secondary affections, we may rank some degree of vasomotor paralysis. It may be doubted if neuralgia ever reaches more than a very slight degree without involving more or less of this; for so-called points douloureux are themselves pretty certainly, for the most part, a phenomenon of vaso-motor palsy; and the more widely-diffused soreness, such as remains in the scalp, for instance, after attacks of pain, even at an earlier stage of trigeminal neuralgia than that in which permanenently tender points are formed, is probably entirely due to a temporary skin-congestion. The phenomenon presents itself in a much more striking way in the condition of the conjunctiva seen in intense attacks of neuralgia affecting the ocular and peri-ocular branches of the fifth; one sometimes finds the whole conjunctiva deeply crimson; and, in one remarkable instance that I observed, the same shade of intense red colored the mucous membrane of the nostril of the same side. In several instances, I have seen a more than usually violent attack of sciatic pain followed by the development of a pale, rosy blush over the thinner parts of the skin of the leg, especially of the calf, which were then extremely tender, in a diffuse manner, for some time after spontaneous pain had ceased.

2. Not merely the circulation, however, but the nutrition of tissues, becomes positively affected, in a considerable number

of cases. It is difficult to judge, with any exactness, in what proportion of neuralgic cases this occurs, but its slighter degrees must be very common. It has very frequently happened to me, quite accidentally, in examining with some care the fixed painful points, which are so important in diagnosis, to be struck with the decided evidence to the finger of solid thickening, evidently dependent on hypertrophic development of tissue-elements; in severe and long-standing cases, I believe this condition will always be found. Probably the change is, more usually than not, sub-inflammatory; but it is certain, on the other hand, that there are great variations in the kind of tissue-changes complicating neuralgia, and that inflammation is no necessary element in them. This subject has greatly engaged my attention, and I find myself able to give what is probably a fuller account of the matter than any yet published connectedly.

The following tissues have been seen by myself to become altered under the influence of neuralgia in nerves distributed to them, or to the parts in their immediate neighborhood.

(a) The hair has changed in color in many cases. Of twenty-seven patients suffering from neuralgia of the ophthalmic division of the fifth, eleven had more or less decided localized grayness of hair on that side. The amount of this varied greatly, from mere patches of gray near the roots of the hair to decided grayness of the majority of the hairs over the larger part of half the head, nearly to the vertex; but in each case it was a change of color that did not exist on the other side of the head. In four of these cases there was also grayness of part of the eyebrow on the affected side. A very remarkable phenomenon, which I have sometimes identified, is fluctuation of the color, the grayness notably increasing during, and for some time after, an acute attack of pain, and the same hairs returning afterward more or less to their original color. My attention was first called to this curious occurrence in my own case. I have so often related this case [see, for instance, my article on Neuralgia in "Reynolds's System of Medicine," vol. ii.] that I shall merely recall the fact that, when pain attacks me severely, the hair of the eyebrow on the affected side displays a very distinct patch of gray (on some occasions it has been quite white) opposite the tissue of the supra-orbital nerve, and that the same hairs (which can be easily identified) return almost to the natural color when I am free from neuralgia. I must, however, add the very curious fact, which I observed accidentally in experimenting (as regards urinary elimination) on the effects of large doses of alcohol, that a dose sufficiently large to produce uncomfortably narcotic effects invariably caused the same temporary change of color in the hair of the same eyebrow, even when no decided pain was produced, but only general malaise.

The subject will be again referred to under the heading of Pathology.

Change in the size and texture of the hairs, in neuralgia, has been noted by Romberg and Notta, and has been several times observed by myself. Occasionally the individual hairs near the distribution of the painful nerve becomes coarsely hypertrophied; at times the number of hairs appears to multiply, but I imagine this is only a case of more rapid and exuberant development of hairs that would be otherwise weak and small. In one very remarkable instance of sciatica this came under my observation; the whole front of the painful leg, from the knee nearly to the ankle, became clothed, in the course of about six months, with a dense fell of hair, which strongly reminded me of similar abnormal hair-growths that have been occasionally seen in connection with traumatic injuries to the spinal cord. More commonly, the effect of neuralgia upon hair is to make it brittle, and to cause it to fall out in considerable quantities; one young lady, who consulted me for a severe migraine, was seriously afraid of having a good head of hair completely ruined in this way, but the hair gradually grew again after the neuralgia had disappeared.

(b) The periosteum of bone and the fibrous fasciæ in the neighborhood of the painful points of neuralgic nerves not unfrequently take on a condition of sub-acute inflammation, with marked thickening and tenderness on pressure. The most striking instance of this that I have seen was in a lady suffering from severe cervico-brachial neuralgia. In the neighborhood of the emergence of the musculo spiral nerve at the outer side of the arm, there was developed what looked for all the world like a large syphilitic node, except that the skin was brightly reddened over it; this disappeared altogether some little time after the neuralgia had been relieved by ordinary treatment. I must say that, but for the peculiar circumstances of the case, putting syphilis out of the question, I could not have avoided the suspicion, at first, that the swelling was specific. But I have several times seen similar, though less developed, swellings in neuralgia, and in one case I noticed the occurrence of such a swelling on the malar bone, in an old woman in whom the neuralgic pain was limited to the auriculo-temporal and the supra-orbital branches of the fifth.

A very important point is to be noted in connection with these sub-inflammatory swellings in connection with neuralgia. Pressure on them will, frequently, not merely excite the neuralgic pains in the branches of the affected nerve, but send a powerful reflex influence through the cord to distant organs, causing vomiting, for instance, or affecting the action of the heart in a very perceptible manner. I shall show, when I come to speak of the phenomena of so-called spinal irritation, that this circumstance has led to erroneous influences in many cases. These exquisitely tender points are often found where Trousseau places his neuralgic *point apophysaire*, namely, over, or very near, the spinous processes of the vertebræ. The tenderness is quite unlike that which is known as hysterical hyperæsthesia; it is much severer, and is limited to one, two, or three points, corresponding, in fact, to the superficial part of the posterior branches of as many spinal nerves.

(c) The nutrition of the skin over neuralgic nerves is sometimes notably affected even when the process does not reach the truly inflammatory stage, which will be more particularly mentioned presently. A certain coarseness of texture of the skin has struck me much, in several cases of long-standing facial neuralgia. And there is a most curious phenomenon (which will be especially considered hereafter in regard to the singular influence of the constant galvanic current upon it), the distribution of a greater or less amount of dark pigment to the skin near the painful part. This phenomenon is much more marked during the paroxysms, and in the slighter cases entirely disappears in the intervals, but in old-standing severe cases it becomes more or less permanent.

(d) The mucous membranes, in situations where we can observe them, not unfrequently show interesting changes, the nutrition of the epithelium of parts covering the painful nerve being exaggerated. It has been noted by various observers, in neuralgia affecting the second and third divisions of the trigeminus, that the half of the tongue corresponding to the painful nerve was covered with a dense fur. This is by no means universally the case, but I have seen it occur several times. In my own case, in which the neuralgia is limited for the most part to the opthalmic division, and only rarely spreads even to the second division of the nerve, this does not usually occur, but I have noticed it on one or two occasions. And I once made the still more singular observation that a large narcotic dose of alcohol, which was sufficient to cause comparatively free elimination of unchanged alcohol in the urine, caused furring of the tongue, which was decidedly thicker on the side of the affected nerve than on the other half of the tongue.

(e) We come now to a group of complications of neuralgia which are exceedingly important, and by no means adequately appreciated as yet, viz., the acute inflammations which directly result from neuralgic affections in a certain percentage of cases, probably much larger than has been at all generally suspected.

The most familiar of the inflammatory complications of neuralgia is herpes zoster, the favorite seat of which is the skin which covers one or more of the intercostal spaces : the eruption, as occurring in this situation, is so well known that it would be waste of time to describe it. In young subjects zoster is commonly painless, at least the sensations are those of heat, pricking, and irritation, rather than of acute pain ; but from puberty onward there is an increasing tendency, especially in those otherwise predisposed to neuralgia, for zoster to be preceded, accompanied, or followed by neuralgia of the intercostal nerves corresponding to the distribution of the eruption. Most commonly, the eruptive period is, in my experience, nearly or quite free from neuralgia, but it often recurs, or breaks out for the first time, when the vesicles are drying up, but more especially if, as is sometimes the case, especially in elderly people, the scabs fall off and leave superficial ulcers. Neuralgia may last, after herpes zoster, for any time from a few days to many weeks, and I have known it so agonizingly severe and so persistent as actually to kill an aged woman from sheer exhaustion. In spite of sundry objections that have been raised to the theory of the nervous origin of zoster, it appears to me that the evidence in favor of it is overwhelming, more especially now that it is proved that the disease, with all the same characteristics presented by it when seen on the chest or abdomen, may occur on the face (following the branches of the trigeminus), or on the forearm (following the course of nerves from the brachial plexus). Two of the severest cases of neuralgia attending herpes that I have ever seen were in private patients (whose family history, unfortunately, I had no means of ascertaining) who were affected, respectively, in the facial and in the brachial nerve-territories.

A far more formidable occasional complication of neuralgia is inflammation affecting the eye. Mr. Jonathan Hutchinson records several cases in which neuralgic herpes zoster of the face was attended with iritis, with serious or even irremediable damage to the organ. For my own part, I have witnessed several instances in which neuralgia of the first and second divisions of the fifth has been attended with skin-inflammation, but only in one of these (just alluded to) did the inflammation present the characteristic appearances of herpes: in all the rest it far more closely resembled erysipelas. The skin was excessively reddened in an almost or quite continuous patch over the whole territory through which ran the painful nerves; by no means only linearly in the course of the nerves, though accurately limited to the district of the first or first and second divisions of the fifth. In the first case I saw (a woman, aged thirty-two), nothing could be more startling than the rapidity with which an irregular patch of the skin, including half of one cheek, the side of the nose, and a large part of the forehead and scalp on the same side, became converted into the dense, fiery-red, brawny tissue, with minute vesicles scattered over its surface, which looks so characteristic of erysipelas ; this commenced immediately on the subsidence of severe neuralgic pain. During the erysipelatoid inflammation, though there was no spontaneous pain, the neuralgia could be instantly lighted up for a moment by pressure on the infra-orbital foramen, on the supra-orbital notch, or upon the malar bone, about its centre. Since that time I have seen several cases of a similar character; two of these, which were reported in the *Lancet* for 1866, I shall here reproduce: [Extensive inquiries convinced me that the tendency to erysipelatous complication of facial neuralgia is exceedingly common. Eulenburg expressly confirms my original statement to this effect, and extends it to all neuralgias.]

CASE I.—A woman, aged sixty-three, presented herself in the out-patient room at Westminster Hospital, suffering from neuralgia of ten days' standing (which for the present, however, seemed to have abated considerably), but asking advice chiefly for an erysipelatoid inflammation which had come on a day or two before, and occupied the area of the painful nerve-district. The neuralgia had affected the supra-orbital nerve, running up toward the vortex, and the auriculo-temporal branch of the third division of the fifth; although there was no very acute pain present at this time, pressure over the supra-orbital notch, or at a point just in front of the ear, would at once cause a brief paroxysm of pain. It was curious to find that there was a thickened and tender spot over the malar bone (and corresponding to the exit of some nerve filaments from the bone) which had never been the seat of spontaneous neuralgia, but pressure here sent a dart of pain into the auriculo-temporal and supra-orbital nerves. The inflammation was markedly limited to the general area of distribution of the twigs of the auriculo-temporal and of the ophthalmic division; it was of a continuous deep-red color, and attended with much thickening of the skin. The conjunctiva was intensely congested, and there were lachrymation and very marked photophobia, but there were no signs of iritis, and no corneal clouding.

CASE II.—M. W., a woman, aged forty-two, well-nourished and healthy-looking, married and had one child; had never suffered any serious ailment except once, about five years previously. She then had a decided attack of "erysipelas," very accurately limited to the right half of the face. Five months before coming to me she sustained a severe shock from being thrown out of a chaise, without suffering any external or visible damage. An hysterical tendency, which she had always possessed, became more marked; it revealed itself by palpitations, occasional dysphagia, and a disposition to weep causelessly. The menses were flowing at the time of the accident; they ceased abruptly soon after (they had been scanty for some time previously), and did not recur till four months later. The hysteric disturbance progressively increased during a fortnight, and then the patient was attacked with violent intermittent neuralgia, commencing in the eyeball and spreading over the district supplied by the branches of the first and second divisions of the trigeminus. The pain was accompanied by intense conjunctival congestion and photophobia [Dr. Handfield Jones remarks that photophobia, in his experience, is only a rare accompaniment of facial neuralgia. I have latterly come to the same opinion. Redness of the eye and lachrymation are very common; true photophobia uncommon. Notta's experience would seem to have been similar]. It lasted on the first day fourteen hours, and returned daily for the next fifteen or sixteen days. An attack of erysipelas, strictly limited to the district of the painful nervous branches, then set in. From that moment the neuralgic attacks became less frequent and severe. A second similar onset of erysipelas occurred some three or four weeks after the first. Finally, the neuralgia disappeared about four months after its first occurrence, and the menses reappeared in tolerable abundance about the same time. About a fortnight before this the patient had discovered that her right eye was dim; as the photophobia had previously disabled her from opening the eye, she could not be sure how long this dimness had existed. At the time of her visit to me the cornea was blurred with a large patch of interstitial lymph, with the remains of a superficial ulcer in the centre; the iris was turbid and discolored, showing the traces of recent but past iritis; the pupil was regular in form and active to light; the conjunctiva was slightly congested. Ophthalmoscopic observation was attempted by a skilled observer, but could not be satisfactorily carried out, from the turbid state of the media. The conjunctiva was slightly congested. In place of the lachrymation that had prevailed during the neuralgic period, there was a remarkable insensibility of the lachrymal apparatus, for the patient had noticed that the smell of onions, which would make the other eye weep profusely, had no influence on the affected one.

The family history of this patient is a most remarkable one. All the members of her mother's family, for two generations back, had died at middle age, either from apoplexy or some disease involving hemiplegia. This case has, by a mistake, not been added to the list of twenty-two private cases in which the family history was carefully investigated, that will be found in the chapter on Pathology; this arose from the fact that the patient was not properly under my care, but was sent to me as a medical curiosity; the notes of her case were therefore taken in a different book from the others. The case certainly ought to be taken as a counterpoise to such a one as No. XVI. in the list, which is that of a gentleman who suffered from the most complicated neurotic maladies (asthma, angina pectoris, facial neuralgia, more than once attended with erysiyelas), but whose family history, so far as it was known, presented no traces of tendency to neurotic disease.

COMPLICATIONS OF NEURALGIA.

To these two cases of inflammation, secondary to neuralgia, I shall add a third, which is even more interesting, and which came under my notice not long since.

CASE III.—H. T., watchmaker's assistant, aged forty-two, suffered for about three weeks with very severe remittent abdominal pain, entirely unconnected with dyspepsia, constipation, or diarrhœa. It was intermittent in character, but observation soon showed that the times at which it came on were simply those at which the stomach had gone longest without food, especially the early morning, and that nourishment never failed to relieve it. The suffering was great, and the man failed considerably in general health, notwithstanding that his appetite and digestion were unimpaired. He had only been under my care about ten days when he presented himself one day at the hospital, and stated that the pains in the stomach had entirely left him, but that he suffered the most frightful pains in and around the right eye. I found a well-marked conjunctival congestion and lachrymation, but there were as yet no tender points; the neuralgia was felt most severely in the globe of the eye and in one tolerably straight tine, darting up toward the vertex from the brow. The iris seemed clear and free, and the cornea was not cloudy. I gave the man a subcutaneous injection of one-sixth grain acetate of morphia, for present ease, and ordered him muriate of iron and small doses of strychnia three times a day. When he next appeared, four days later, I was alarmed to perceive that unmistakable iritis had fully developed itself, the iris was already turbid and discolored and the pupil irregular, from a serious amount of adhesions. By this time there were fully-developed tender points, supra-orbital and parietal; besides this, pressure on the globe caused paroxysms of pain in all the branches of the ophthalmic division, but there was not much spontaneous pain. I dropped atropine in the eye, applied blistering fluid to the back of the neck, [the nape of the neck is the point most suitable for blistering which is intended to affect the eye, and the ophthalmic division of the fifth, generally,] and desired the man to come to see me at my own house next day, intending to take him to an ophthalmic surgeon. Unfortunately he failed to do this, and three days later, when he came to see me at the hospital, the cornea was studded with opacities, the pupil was almost closed with effused lymph, there was violent ocular pain, and a great and increasng sense of tension. I begged him to go without loss of time to the Eye Hospital, as my own ophthalmic colleague was not at Wesminister that day; and I have never heard any more of the patient.

Glaucoma is a still more serious disease of the eye, which I think there is now sufficient evidence to show is sometimes entirely, and very often in considerable part, neuralgic in its origin. Since my attention was directed, some six years ago, to the frequent connection between the so-called rheumatic iritis and neuralgia, I have taken much interest in the subject of acute eye-affections; and the occurrence of one or two cases of glaucoma in personal friends of my own has made this interest even painfully strong. I am necessarily without the means of personally observing glaucoma on the large scale, but I have now seen two cases in which, if I possess any faculty of clinical observation whatever, the whole genesis of the disease was a neuralgic disorder of the trigeminus; and it was to me a melancholy reflection that nothing better than iridectomy in one case, and excision of the eyeball in the other, could be done in the present state of ophthalmic science. There are now a good many recorded instances of neuralgic glaucoma, and Mr. R. Brudenell Carter, of St. George's, and the South London Ophthalmic Hospital, recently assured me that nervous aspect of some form of glaucoma presents itself the strongly to his mind, though he does not commit himself to any theory. Two cases were reported by Mr. Hutchinson, in Ophthalmic Hospital Reports IV. and V.; but the most complete and interesting cases that I have met with are recorded by Dr. Wegner;* they are two out of four that occurred within a very short time in the clinic of Prof. Horner at Zurich, and they form the basis of some researches by Wegner into the nature of the influence of the trigeminus upon ocular tension, which will be referred to, along with others, in the chapter on Pathology. The second of these cases is so important that I shall reproduce it in full.

A. Hediger, aged twenty-four, a moderately strongly-built young woman, seen first in August, 1860. From her own and her mother's account, it seemed she had long suffered from convulsive attacks that did not appear to have been truly epileptic. Some days previously her left eye became very painful, and the sight failed, without any inflammatory symptoms. On inspection the pupil was somewhat dilated, the eye somewhat hypermetropic, fundus normal; No. 5, Jager's type, was read with difficulty. Wegner could not explain the condition. At the end of October the eye was much worse; after severe paroxysms of pain, No. 16 type was the smallest legible, the field of vision was decidedly limited in all directions, but especially on the inner and upper portions. An unusually long hysteric attack was now observed. The patient was for twenty-four hours in a half-sleep, the extremities, meantime, were much jerked, the speech sometimes coherent and sometimes incoherent; she cried out to her friends, etc., but had no severe convulsion-fit with spasm of glottis. She was removed to the hospital, where she stayed six weeks. The hysteria improved under treatment with valerien

---

*Archiv fur Ophthalmologie, B. xii,, Abth. 1, 1866.

and morphia (Prof. Greisinger had confirmed the opinion that there was no true lesion of the centres), but the neuralgia of the globe was extraordinarily severe, both day and night. From January to June, 1861, Wegner saw her occasionally. The visual power of the left eye fluctuated between 15 and 19 Jager. Field of vision very limited. Pupil very dilated and insensitive, the globe painful to the touch, and injected. The right eye weakly hypermetropic; normal field of vision, normal pupil, no pain. The scene suddenly changed on the 29th of June. She was attacked with fearful pain, and an enormous mydriasis with extreme amblyopia of the right eye; the fingers could hardly be counted when placed quite close. The optic disc appeared somewhat cloudy, with very evident venous pulsation. The mydriasis, amblyopia, and neuralgia lasted some time, while simultaneously the left eye could only read 19–17 type, but was painless. The pathology seemed quite obscure, and the surgeon remained almost passive till August, when he performed paracentesis on the left eye. The patient could distinguish fingers at that time at a foot's distance with the right eye; with the left read No. 11, but suffered fearful pains. These diminished after the puncture; the eye could read No. 20 next day, and improved after that to 19; the pains recurred in the next day, but for the first time ceased to disturb sleep. The scene again changed in the most surprising manner on the 27th of August. The most frightful pain again attacked the left eye. The pupil was dilated to the maximum (far beyond what occurs in oculo-motor paralysis); the globe was extremely painful on touch, visual power fallen to 19 Jager. On the other hand, the right eye had a normal pupil, was painless, and could read No. 12. Parencentesis of the left eye improved its vision and diminished pain, but only temporarily, so that it had to be repeated at short intervals. The condition was so far stationary toward the end of October that the right eye continually gained visual power, but the left stood still and fluctuated from worse to better, with the greater or less severity of the neuralgic paroxysms. Pupils always in extreme dilatation. In the end of October and beginning of November (the patient had worn a large seton for a month) remarkable changes occurred; the neuralgia of the left globe diminished steadily, the pupil got smaller, the visual power increased, the neuralgia now was only on the lower lid, which was slightly red and painful to the touch, and had continual spontaneous pain. Visual power of right eye No. 3, of left eye No. 5. Visual field intact; with full illumination by weak light there is a peripheral torpor, but only in a narrow zone. The hyperæmia now extended more and more over the lower lid and the upper part of the cheek; this was apparent during the paroxysms, which were very severe, and destroyed sleep; it did not allow the skin to be touched; the color was deep (with high temperature) and extended to the angle of the mouth. This phenomenon lasted till the beginning of December, when neuralgia again attacked the left globe, with strong mydriasis and diminution of visual power (15 to 20 Jager), till at last the movements of the hand could hardly be distinguished, and this state of things continued with fluctuations up to the end of the month. The seton had been taken off just before the new outbreak; it was put in again on December 31st. In January the pains continued severe in the eye, with only one remission (from the 17th to the 20th), when the hyperæmia recurred in the cheek. On the 26th the pupil was very dilated, and fingers could not be seen at half a foot's distance. Visual field very limited, globe hard. A large upper iridectomy was made. After this the pupil was contracted, the pains diminished, visual power 10 Jager, field seven inches. In the middle of February the hysterical attacks recurred with great force; the patient was unconscious half the day; she was clear enough in senses when awake, but complained of buzzing in her head, as if a cock-chafer were inside it. From this till the middle of March, the left eye did not alter, the impairment of vision remained, with normal pupil and no pain in the globe, and the iridectomy seemed at least to have done good in one direction; but on the 13th of March the operated eye was again attacked with pain, visual power fell to No. 17, pupil became dilated, and ofter a few days the swelling, heat, and tenderness of the cheek recurred. During the years 1862 and 1863 the condition remained pretty much the same; *i. e.*, the right eye sound, the left painful (in spite of the iridectomy) with dilated pupil, concentrically narrowed visual field, visual power fluctuating between No. 15 and mere finger-counting without any opthalmoscopic appearances. A number of paracentesis and subcutaneous injections of morphia (which last were the more indicated as the supra-orbitalis was tender on pressure) always brought relief merely for a few hours. On the 19th of April, 1864, vision being complete in right eye, and No. 19 in left, Wegner punctured the latter. On the 2d of May the eye read No. 10 slowly, the pains had gone and not returned, the pupil became smaller. On the 31st of March, 1865, the patient was pronounced well; the eye was painless, the pupil somewhat larger than the other; the finest type could be read when looked at very close.

3. The next group of affections secondary to neuralgia are the paralysis of muscles. These are pretty common; I find them in twenty-eight of the hundred cases which have been referred to. But of these twenty-eight instances of paralytic affections no less than twelve were connected with neuralgia of the trigeminus, and in most of these it was one or more of the muscles connected with the eye that were affected. Sciatica is nearly always attended with much weakening of vol-

untary power of the muscles of the thigh and leg; and in some instances this reaches to decided or even complete paralysis. In looking for this phenomenon we must be very careful that we do not mistake the mere reluctance to move the limb, on account of the painfulness of all movements, for true paralytic weakness of nerve and muscle. And it is also necessary to bear in mind, in prolonged cases, the probability that much of the weakness may have been caused by degeneration of the muscles owing to forced inaction. Still, there is a class of secondary paralyses that are in no way to be confounded with such effects as these: for instance, it occasionally happens, almost in the very first onset of severe sciatic pain, that the limb hangs absolutely helpless; and in one such case lately, being struck with the completeness of the loss of power, I tested the Faradic irritability by directing a sharp current on comparatively exposed portions of the painful nerve (*e. g.*, in the popliteal space, and behind the head of the fibula), and elicited only the most feeble contractions, entirely unlike what the same current evoked in the opposite limb. I regret that I have as yet found it impossible to carry out a regular inquiry as to the sensibility to the different currents of motor nerves which are centrally connected with neuralgic sensory nerves.

Muscular viscera which are composed of unstriped fibre, like the intestines, or of a mixture of striped and unstriped, like the heart, are probably very liable to a secondary paralytic influence from certain special neuralgiæ. It is ascertained that the pain of a certain degree of severity in the branches of the fifth may absolutely stop the heart's action for a moment—an effect which is succeeded, usually, by violent and disorderly pulsations. I have myself once known the operation of "pivoting" a tooth, which gave frightful pain, cause instantaneous and most alarming arrest of the heart's motion, which for a minute or two seemed as if it were going to be fatal. But the variety of visceral paralysis which is probably far the most frequent is secondary paralysis of the bladder, from neuralgia in one or other of the pelvic organs, or of the external genitalia; and next to this comes paralytic distension of the cæcum, colon, or rectum, secondary to various abdominal and pelvic neuralgic affections. In one instance of acute ovarian neuralgia that I saw, the paralytic distention of the colon was by far the most remarkable circumstance, so enormously was it developed; and for some days after the neuralgia had ceased, and when the flatulence had nearly disappeared, the intestine remained absolutely torpid.

4. Convulsive actions of muscles, as every one knows, are very common complications of neuralgia. In trigeminal neuralgias these may be observed (according to the division or divisions of the nerve that are affected) in the proper muscles of the eye, or in those supplied by the fourth and sixth nerves, or

(perhaps only when two or three divisions of the fifth are neuralgic at once) by the portio dura. It is curious, however, that those formidable spasmodic affections of the face which belong to the same order as torticollis and writer's cramp, are not frequently, if ever, directly associated with trigeminal neuralgia. The only connection between them seems to be that these peculiar spasmodic affections are only developed in highly-neurotic families, some of whose members are almost sure to be found suffering from some form of regular neuralgia. In severe sciatica it has several times happened to me to see convulsive action of the flexors, bending the leg spasmodically upon the thigh. And in a very large proportion of all neuralgias, wherever situated, attentive observation of the patient during the paroxysms will detect the existence of local twitching or local spasm of muscles, though these may be slight in degree.

Among the convulsive affections must be reckoned convulsive movements and tonic spasms of various portions of the alimentary canal. Vomiting is a common example of this; in migraine it is the regular and necessary climax of attacks which last with severity for a certain time; indeed, any severe attack of neuralgia involving the ophthalmic division of the fifth may excite vomiting. Convulsive action of the pharyngeal muscles, as a complication of pharyngeal or laryngeal neuralgia, occasionally occurs to such an extent as to render deglutition difficult or impossible for the time. And I have seen what I do not doubt to have been a spasmodic condition of the rectum induced by peri-uterine neuralgia. The genito-urinary organs are also not unfrequently affected spasmodically in consequence of a neuralgic affection either peri-uterine or pudendal. I have seen spasmodic stricture of the male urethra thus produced, and likewise vaginal spasm.

5. Impairments of sensation, both common and special, are very frequent attendants of neuralgia. As regards the special sensations, we may first mention that of touch; this is almost constantly impaired, immediately before, during, and some little time after a neuralgic paroxysm, in the skin supplied by the painful nerves. I was first led to make this observation by my own experience; the skin all round the inner angle of my right eye is permanently less sensitive to distinctive impressions than that of the opposite side, and this impairment is always decidedly greater, and spreads over a larger surface, before, during, and for some time after, the attacks of pain. More extended observation has convinced me that a certain amount of bluntness of distinctive skin-sensation accompanies nearly every neuralgia. As regards the sense of taste, I have found this decidedly perverted, at the time of an attack, even in my own case, although the neuralgia never extends into the third branch of the nerve. It is interesting to notice, in connection

with this, that the epithelium of my tongue has been seen, on one occasion, to be exaggerated on the side of the neuralgic affection, showing a probability that there is perturbed function, at any rate of certain fibres, of the third division. But I have seen much more decided alteration, indeed temporary entire abeyance of the power to distinguish between the tastes of different substances, with the affected side of the tongue, in a case of severe epileptiform tic in which the third division was strongly affected with neuralgia; and Notta records a similar instance. As regards vision, besides minor perversions and disturbances, I have observed more or less complete amaurosis in several instances of ophthalmic neuralgia; in one case it was absolute, and lasted, with but slight improvement in the intervals between the paroxysms, for nearly a month, but disappeared entirely, though somewhat gradually, after the final cessation of the neuralgia. As regards hearing, I have noticed serious impairment only in five cases, all of them of a severe type of trigeminal neuralgia, involving all three divisions of the nerve. Smell, I have never observed to be more than doubtfully impaired, except in one case (*vide* Chapter III), where it was completely destroyed.

Common sensation was reported by Notta as affected in only three cases out of a hundred and twenty-eight; but my own experience has afforded a much larger proportion of instances in trigeminal neuralgia. Indeed, in all situations neuralgia appears to me to involve this effect, in the larger number of instances, in the early stages; later, it is supplanted in part by great tenderness on pressure in the well known *points douloureux*, and sometimes the tenderness becomes diffused over a considerable surface. I agree with Eulenburg in thinking that anæsthesia is more frequent in sciatica than in other neuralgias.

6. Secretion is often very notably affected in neuralgia; the phenomena are necessarily more easily observed in connection with affections of the trigeminal than of other nerves. In the great majority of cases the affection is in the direction of increase; at least, the watery elements of secretion are often poured out in profusion. Thus, profuse lachrymation is exceedingly common in ophthalmic neuralgia; in a large number of cases there is also copious thin nasal flux on the affected side; sometimes, however, the secretion, though copious, is semipurulent, or bloody. Increased salivation has been noticed, by a large number of observers, in neuralgia involving the lower division of the fifth. In a smaller number of instances, the secondary effect on secretion is precisely opposite; thus both Notta and myself have observed complete dryness on the nostril on the affected side in ophthalmic neuralgia.

I might expand this chapter on the complications of neuralgia to a very much greater length; but, as regards the clinical history of these affections, it is perhaps better not to occupy more time and space. It will, however, be necessary to return to the consideration of the subject in connection with Pathology.

CHAPTER III.

PATHOLOGY AND ETIOLOGY OF NEURALGIA.

The pathology and the etiology of neuralgia cannot be considered apart; they nust be discussed together at every step. I do not mean to say that neuralgia is singular among diseases in this respect; it seems to me merely a case in which the intrinsic defects of the conventional system of separating the "causes" of disease from its pathology happen to be more glaring and more easily demonstrable than usual.

Neuralgia possesses no "pathology," if by that word we intend to signify the knowledge of definite anatomical changes always associated with the disease, in a manner that we can exhibit or exactly describe. It also possesses no demonstrable causes, if we employ the word "causes" in the old metaphysical sense. And yet I am very far from admitting, what seems to be so generally taken for granted, that we know less about the seat, the nature, and the conditions of neuralgia than of other diseases. On the contrary, I believe, with all deference to the supporters of the ordinary opinion, that we know more about neuralgia, in all these respects, than we do about pneumonia, only our knowledge is not of the superficial and obvious kind, but requires the aid of reason and reflection to develop and turn it to account. It has long been a matter of surprise to me, that even able writers have been content to talk about this disease (as, indeed, they have been content to speak of many nervous diseases) with an inexplicable looseness of phraseology. They speak of its "protean" forms; whereas, in my humble judgment, its forms are by no means specially numerous. They insist on the mysterious and unintelligible manner of its outbreaks, remissions and departure; but I shall try to show that, although, in the investigation of neralgi*, we are continually stopped in particular lines of inquiry by what seems to be ultimate facts, susceptible of no further immediate solution, the channels of information open to us are so unusually numerous as to enable us to accumulate a mass of information which, upon further reflection, will be found to furnish the materials of a synthesis of the disease singularly clear and effective for every practical purpose of the physician. In one important particular I especially hope to convince the reader that a large proportion of the mystification as to the pathology of neuralgia is gratuitous, and the result of great carelessness in estimating the comparative value of different facts. I hope to show clearly that, as regards both the seat of what must be the essential part of the morbid process, and the general nature of the process itself, we possess very definite information indeed. I expect, in short, to convince most readers that the essential seat of every true neuralgia is the posterior root of the spinal nerve in which the pain is felt, and that the essential condition of the tissue of that nerve-root is atrophy, which is usually non-inflammatory in origin. This doctrine seems, at first sight, presumptuous,* in the confessed absence or extreme scarcity of dissections which even bear at all upon the question. But one source of the extraordinary interest which the pathology of neuralgia has long possessed for me resides in this very fact, that I am convinced we can demonstrate the above apparently difficult theorem by means of pathological observations on the living subject, taken in conjunction with physiological experiments, and with only the aid of a very few isolated facts of positive morbid anatomy. I need hardly say that I am none the less anxious for that further assurance which we shall one day, perhaps, obtain by means of greatly-improved processes for microscopic detection of minute changes in nerve-centres; but, looking to the necessary rarity of opportunities for post-mortem examinations of the nervous system in any but the most advanced stages of neuralgias, it will hardly be disputed that, if I am right in my main position, we are singularly fortunate to be so unusually independent of the need for this source of information.

1. The first fact which strikes me as of decided importance is the position of neuralgia as an hereditary neurosis; and this character of the disease is so pregnant with significance, that I shall take some considerable pains to put the fact beyond doubt in the reader's mind.

There are two series of facts which support the theory of the inheritance of the neuralgic tendency: (a) instances in which the parent of the sufferer had also been affected with the disease; and (b) instances in which the family history of the patient being traced out more at large it appeared that, among the members of two or more generations, while one, two, or more individuals had been actually neuralgic, other members had suffered from other serious neuroses (such as insanity,

---

* Eulenburg, to whose excellent work ("Lehrbuch der functionellen Nerven-krankheiten," Berlin, 1871) I shall have frequent occasion to refer, has partly misunderstood the drift and scope of my argument, a misfortune which I owe to the impossibility of giving, in the "system of Medicine," more than the briefest and most superficial sketch, both of my ideas and of the facts on which they rest.

epilepsy, paralysis, chorea, and the tendency to uncontrollable alcoholic excesses), and, in many instances, that this neurotic disposition was complicated with a tendency to phthisis.

(a) The question of the direct transmission of neuralgia itself from the parent seems the easiest of decision, though even this cannot always be satisfactorily cleared up by the hospital patients, among whom one collects the largest part of one's clinical materials. However, I have been at the pains of investigate a hundred cases of all kinds of neuralgia, seen in hospital and private practice, with the following results: twenty-four gave distinct evidence that one or other parent had suffered from some variety of neuralgia; fifty-eight gave a distinctly negative answer; and eighteen would not undertake to give any answer at all. Among the twenty-four affirmatives are inserted none in which the history of the parent's affection did not clearly specify the liability to localized pain, of intermitting type, but recurring always in the same situation during the same illness. In three of these twenty-four instances, the patient stated that both parents had suffered from such attacks, and, in one of these, it appeared that the grandfather had likewise suffered,

(b) The question of the tendency of a family, during two or more generations, to severe neuroses of more or less varying kinds, including neuralgia, is difficult to work out perfectly, though in a large number of instances we may get enough information to be very useful. I have spent much time and trouble in endeavoring to collect such information; but there are two main difficulties in connection with all such attempts. From hospital patients you frequently can get no reliable information whatever respecting any members of the family farther back than the immediate parents; and, even respecting uncles and aunts and first cousins, it is often impossible to learn any thing. And when you get to a higher class of society, especially when you approach the highest, although the information may exist, it may be withheld, or you may be purposely mystified. One would doubt beforehand, under these circumstances of difficulty, whether it would be possible to obtain affirmative evidence of the neurotic temperament of the families of neuralgic patients in general; but, in truth, the evidence is so overwhelming in amount, that more than enough can be obtained for our purpose. I shall give, first, the results of one special inquiry which, by the kindness of a patient, I have been able to carry out with more than usual completeness; it relates to the medical genealogy of a sufferer from sciatica; the account is fairly complete for four generations. The great-grandfather was a man of splendid physique (an only son), who lived very freely, but died an old man. His children were three sons, one of whom (though strictly temperate) was a man of eccentric and somewhat violent temper, and suffered from a spasmodic facial affection. This one, the grandfather of my patient, married a lady who died of phthisis, and among the ten children she bore him, two sons died of phthisis, two sons became chronically insane, one son died, probably of mesenteric tubercular disease (aged fifty-six), two sons are still alive at very advanced ages, and have always been perfectly healthy and strong; one daughter died in middle age, it is not certain from what cause; one daughter lived healthily to the age of eighty, and then was attacked by facial erysipelas, followed by violent and intractible epileptiform tic, which clung to her for the remaining four years of her life; and the remaining daughter, an occasional sufferer from migraine, died at the age of sixty-seven, almost accidentally, from exhausting summer diarrhœa. The fourth generation, in this branch of the family, consisted of thirty-one individuals; of whom seven have died of phthisis, or scrofulous disease; one from accidental violence, one from rheumatic fever, one from scarlet fever; and among the surviving twenty-two one has been insane, but recovered; two are decided neuralgics; one is occasionally migraineuse, and once had a smart attack of facial erysipelas, corneitis, and iritis, as the climax to a severe neuralgic attack; one has been a sufferer from chorea; one has become phthisical; one developed strumous disease, but has fairly recovered from it. The remaining fifteen enjoy good health, but are distinguished, almost without exception, by a markedly neurotic temperament, indicated by an anxious tendency of mind, quickness of perception, æsthetic taste, disposition to alternations of impulse and procrastination. Of the young fifth generation growing up, there have been twenty-five children, of whom only one has died (from fever), the rest are apparently healthy (most of them specially so); but, as few have yet reached the age for the development either of phthisis or of neurotic diseases, the future of this generation can only be guessed at. [It is unnecessary to trace the other descendants of the second generation, but I may state that their medical history, also, strongly supports the theory of inheritance of the neurotic tendency, and of the influence of an imported element of phthisis in aggravating the latter.] I suspect that, as regards the young children now growing up, everything will depend on the care with which they are fed, and the kind of moral influences brought to bear on them, two subjects which will be fully dwelt on in the chapter on Treatment.

Of less perfect inquiries on the subject of neurotic disposition inherited by neuralgic patients, I have made a great number, though I regret to say that I have not attempted the task in the whole number of those from whom I inquired as to direct inheritance of neuralgia from their parents. However, in eighty-three cases this was done with all possible care, and any deficiency of completeness in the results is not my fault. I shall take first those that were private patients, twenty-two in number, respecting whom, I may say, that the evidence is of the best, as far as it goes, since I was better able to discriminate as to the worth of statements, than in dealing with hospital patients, and have rejected every case in which the informant did not seem intelligent enough, or otherwise to have the means, to give a thoroughly reliable account.

I. Neuralgia cervico-brachialis; in a lady, aged seventy-one. Mother suffered suffered from epileptiform facial tic; uncle was paralyzed; patient herself eccentric to the verge of insanity.

II. Bilateral sciatica of great severity; in a gentleman, aged seventy-three. Gout, paralysis, and neuralgia, have been frequent in the family.

III. Cardiac neuralgia; in a man, aged twenty-four. Father epileptic and a drinker; grandfather died of softening of the brain, aged thirty-eight.

IV. "Cerebral" neuralgia; in a single lady, aged thirty-eight. Mother has been insane; first cousin epileptic.

V. Lumbo-abdominal neuralgia; in a gentleman, aged fifty-two. Father a drinker; mother insane; maternal grandfather phthisical.

VI. Severe neurotic angina pectoris; in a gentleman, aged fifty. Almost every one of the graver neuroses among patient's near relations.

VII. Migraine and cervico-occipital ueuralgia; in a young lady, aged twenty-five. Immediate causes, brain-work, and influence of cold weather. Father and brother both epileptic; father's family much affected with neurotic diseases.

VIII. Sciatica; highly-nervous temperament. Father died insane from drink; and probably other members of the family also nearly or quite insane.

IX. Auriculo-temporal neuralgia; in a married lady, aged twenty-eight. Father's family markedly phthisical and neuralgic.

X. Intercostal neuralgia; in a girl (phthisical), aged twenty-four. Mother and two uncles phthisical; maternal grandfather epileptic and a drinker.

XI. Facial neuralgia (third branch trigeminal); in a gentleman, aged fifty-four, a great whiskey-drinker. Drinking hereditary for three generations; father died insane; grandfather epileptic; sister phthisical; two brothers very "eccentric."

XII. Migraine, severe; in a lady, aged thirty-three. Grief was the immediate cause. Mother hemiplegic at forty-second year; first cousin insane; two aunts (maternal) epileptic.

XIII. Extremely severe sciatica and cervico-brachial neuralgia of the left side, with singular inflammatory consequences; in a lady, aged fifty-two. A family history remarkably free from neurotic diseases and from phthisis. The neuralgia was probably caused partly by excessive ptyalism, partly by over brainwork.

XIV. Migraine; in a young lady, aged sixteen; very profuse menstruation, which had lasted for two years. Family history very free both from phthisis and neuroses.

XV. Frontal and nasal neuralgia; in a man. Repeated attacks of localized facial erysipelas; drinking-habits for some years; fatal acute insanity in middle age. Father insane, committed suicide; mother subject of violent epileptiform tic.

XVI. Angina pectoris (neurotic); spasmodic asthma, twenty years; facial neuralgia and erysipelas; in a gentleman, aged fifty. Family medical history scanty and imperfect; but, as far as it goes, entirely without evidence of either phthisis or neuroses.

XVII. Neuralgia of testis, immediately caused by local irritation. Father died of phthisis; paternal uncle epileptic and insane.

XVIII. Ovarian neuralgia; in a girl, aged twenty-six, liable to occasional migraine. Mother has suffered sciatica; brother died of phthisis.

XIX. Gastralgia; in a man, aged twenty-seven; highly intellectual and nervous. Famity history very free from neuroses; but some evidence of phthisis, in two previous generations, on mother's side.

XX. Sciataca; in a lady, aged sixty; second attack. Ancestors, on both sides, for some generations, clever, and in several instances decidedly eccentric, if not insane; much neuralgia in the family.

XXI. Migraine; in a young lady, aged seventeen; menstrual difficulties. No neurotic nor phthisical family history.

XXII. Sciatica; in a married lady, aged twenty-seven; first pregnancy; had rheumatic fever and subsequent chorea in childhood. Paternal uncle epileptic; mother had rheumatic fever and cardiac disease; paternal grandfather suffered from sciatica late in life.

No one, I think, can look down the above list and fail to be struck with the great preponderance of cases in which the general neurotic temperament plainly existed in the patients' families; and let me add that, in not a few of these cases, the neuralgia in the individual under observation might have been easily set down as dependent merely upon peripheral irritation, which, indeed, plainly did act as a concurrent cause.

Fortunately, however, I am not dependent upon my own evidence alone, for the proofs of the proposition that neuralgia is eminently a development of hereditary neuroses. The great French alienists, Morel and Moreau of Tours, some years ago laid the foundations of the doctrine of hereditary neurosis. They enforced this chiefly with reference to the manner in which insanity is transmitted through a chain of variously-neurotic members of a family stock; and Moreau laid special stress on the deeply interesting connection of the phthisical with the neurotic tendency. Since then various observers have insisted on the same thing. Of late, Dr. Maudsley has worked out this subject with great ability, in his work "On the Physiology and Pathology of Mind," and in his recent "Gulstonian Lectures;" and Dr. Blandford dwells on it with emphasis in his interesting "Lectures on Insanity." [Dr. Blandford does not, however, admit that the phthisical diathesis has any such close and causal relation with neuroses as has been imagined by some recent pathologists; and, on the other hand, he points out that phthisis in neurotic subjects, *e. g.*, the insane, must, in a large measure, be considered the product of the accidentally unhealthy circumstances in which they pass their lives. In the latter opinion I entirely agree.] Indeed, it may be taken as a recognized fact, among the more advanced students of nervous diseases, that hereditary neurosis is an important antecedent of neuralgia, in at least a very large number of instances. I shall conclude this part of the argument by stating the general results of my inquiries respecting sixty-one hospital patients. Of these cases, twenty-two were migraine, or some other affection of the ophthalmic division of the fifth nerve; seven were sciatica; two were epileptiform facial tic; ten were neuralgias affecting chiefly the second and third divisions of the fifth nerve; three were intercostal neuralgias pure; one was intercostal neuralgia plus anginoid pain; seven were intercostal neuralgias with zoster; three were brachial neuralgias; and five were abdominal neuralgias (hepatic, gastric, mesenteric, etc.) Of eighty-three hospital and private patients [It must be understood that the respective numbers do not indicate with any accuracy the relative frequency of the different neuralgias as seen in my practice. (Sciatica, *e. g.*, was proportionally more frequent.) They represent but a small part of the neuralgic patients whom I have seen during fourteen years of dispensary, hospital, and private practice, and they were selected for inquiry merely because I happened to be able to give the time for the necessary questions. Every one who knows out-patient practice will understand how seldom this happened.] I obtained evidence of the presence, among blood-relations, of the following diseases: Epilepsy, fourteen cases (eight were examples of migraine); hemiplegia or paraplegia, nine cases; insanity, twelve cases; drunken habits, fourteen cases; "consumption," eighteen cases; "St. Vitus's dance," four cases. I am well aware that these figures must be taken with caution, and that considerable doubt must rest on the accuracy of some of these details, more especially with regard to "epilepsy," as it was impossible, with the greatest care, to be sure that this was not given, by mistake, for hysteria in some cases; and the same may apply to the statement that relations had suffered from "consumption." The facts are given for what they are worth, and with the express reservation that their total reliability is far less than that of the accounts obtained respecting private patients belonging to the more educated classes. But, in one respect, viz., as regards drunken habits, it is possible that a truer estimate is gained from the statements of hospital patients than from those of private patients, who would usually be more prone to reticence on such a topic.

The evidence as to the hereditary character of neuralgia assumes a yet higher importance when supplemented by the facts respecting the alternations of neuralgia with other neuroses as the same individuals. Every practitioner must be aware how frequent is the latter occurrence. Nothing is more common, for example, than to see insanity developed as the climax of minor nervous troubles, especially of neuralgia. And there is one form of neuralgia, the true epileptiform tic, which is intimately bound up with a mental condition of the nature of melancholia, and even with the markedly suicidal form of the latter affection. I have lately had under my care a lady in whom the prodromata of a severe facial neuralgia were mental; the disturbance commenced with frightful dreams, and there was great mental agitation even before the pain broke out; this disturbance of mind, however, continued during the whole period of the neuralgia, and was relieved simultaneously with the cessation of the attacks of pain. This is contrary to what happens in some cases; thus, Dr. Maudsley quotes the case of an able divine who was liable to alternations of neuralgia and insanity, the one affection disappearing when the other prevailed. Dr. Blandford has met with several instances in which neuralgia has been followed by insanity, the pain vanishing during the mental disturbance, and reappearing as the latter passed away. And he remarks that, in the transition of a neuralgia (to mental affection), we may well believe that the neurotic affection is merely changed from one centre to another, from the centres of sensation to those of mind. He says that the ultimate prognosis of such cases is bad; a point to which we shall have to refer again.

The prominent place which quasi-neuralgic pains hold in the earlier history of locomotor ataxy is a fact that cannot but engage attention. In this volume we have not treated these pains as belonging to the truly neuralgic class, for the very practical reason that they are but incidents in a most important organic disease, and that in a diagnostic and prognostic point of view it is necessary to dwell on their connection with that disease. But, in considering the pathological relations of neuralgia, it would be improper to omit the consideration of the pains of locomotor ataxy, which bear a striking semblance to neuralgic pains. The fact that they are an almost if not quite constant feature of a disease which is from first to last an atrophic affection (mainly of the posterior columns of the cord), in which the posterior roots of the nerves are almost always deeply involved, has a bearing on our present inquiry too obvious to need further remark.

Equally important to our investigation is the fact that pains, closely resembling neuralgia, are not very uncommonly a part of the phenomena of commencing, and more frequently of receding, spinal paralysis. I have the notes of three cases of partial recovery from paraplegia, in all of which the patients remained for years, in one case for nearly twenty years (ending with death), the victims to a singularly intractable neuralgia of both lower extremities. In the worst of the cases the patient was the victim of excessive and continuous labor at literary work of a kind which hardly exercised the mental powers, but was extremely exhausting to the general power of the nervous system; he broke down at about the age of fifty, but dragged on a painful existence for the long period above mentioned.

We are also certainly entitled to adduce the example of the so-called neuralgic form of chronic alcoholism as an instance of the close relationship of neuralgia to other central neuroses. I refer to those cases, more common perhaps than is generally admitted, in which pains in the extremities, often quite resembling neuralgia in their intermittence, are either superadded to or take the place of the muscular tremors and general restlessness that are more popularly considered as the essential nervous phenomena of chronic alcoholic poisoning. That the pains are usually bilateral, and more diffuse in their character than those of ordinary neuralgia, is a fact which it is not difficult to explain by the *modus operandi* of the cause; but we shall have more to say on the general relations of alcoholic excess to neuralgia presently. The pains themselves will be fully described in the second part of this book, which treats of the affections that simulate neuralgia; here we need only remark that it is not uncommon for them to occur interchangeably with true neuralgia in the same person.

The occasional interchangeability of migraine with epilepsy is a well-known fact; every practitioner who has seen much of the latter disease will have seen some cases in which the patient had been liable, at some point of his medical history, to "sick-headaches" of a truly neuralgic kind; although it is quite true, as Dr. Reynolds points out, that the kind of sensorial disorder specially premonitory of the attacks consists rather in indefinable distressing sensations, than in actual pain. The aological connection between migraine and epilepsy is, as I have already stated, apparently very close. Such instances as one mentioned by Eulenburg are rightly explained by him; it is the case of a girl who suffered at an unusually early age (nine) from migraine; her mother had·been a migraineuse, and her sister was epileptic; the strong neurotic family tendency is believed by Eulenburg to account for the appearance of migraine at such a period of life.

This seems the fitting place to introduce some special remarks on migraine in its relations to other neuralgias of the head, because Eulenburg has mentioned and combated my view, according to which migraine is a mere variety of neuralgia of the ophthalmic division of the fifth nerve. I call it my view, because, though several other authors had previously expressed it, I was first lead to entertain it by observations made before I had studied their works, and especially by the impressive teaching of my own case, as to which more will be presently said. Eulenburg, though he fully allows that migraine is a neuralgia, urges a series of objections to the identification of migraine with ophthalmic neuralgias; of which objections one, based on the doctrine of Du Bois Reymond as to the action of the sympathetic in migraine, must be reserved for consideration when we discuss the general pathology of the vaso-motor complications of neuralgia. The other grounds of distinction that he urges are the following: In the first place, he remarks that the site of the pain is by far less distinctly referred to definite foci on the outside of the skull than in trigeminal neuralgia; the patient's sensations very usually lead him to declare that the pain is in the brain itself. Secondly, he says that the points douloureux (in Valleix's sense) are almost constantly absent in true migraine. Thirdly, he specifies the character of the pain in migraine—dull, boring, straining, etc.—as differing from that of trigeminal neuralgia, which is ordinarily much more acute and darting. Fourthly, he notes the long duration of individual attacks of migraine, and the long intervals (very commonly three or four weeks) between them. Fifthly, he dwells on the frequent prodromata of migraine referable to the organs of sense (flashes before the eyes, noises in the ears), or to the stomach (nausea), or more generally to the reflex functions of the medulla oblongata (*e. g.*, convulsive rigors, excessive yawning, etc.)

Now, I should have nothing to say against the accuracy of this description, did it apply merely to the distinctions between highly-typical cases of the "sick-headache" of the period of bodily development, and highly-typical cases of the ophthalmic neuralgias which are commonest in the middle and later periods of life; nor indeed should I greatly care if it were finally decided that migraine and clavus should be separated from the true trigeminal neuralgiæ, provided the following points were well impressed on the minds of practitioners. In the first place, I must insist that in my own experience the great majority of undoubtedly neuralgic headaches, which subordinate stomach disturbance, are far less sharply separated than the above description would allow from the unmistakable trigeminal neuralgias; it is only a minority of cases that wear this extreme type, and a far larger number shade imperceptibly away toward the type of ophthalmic neuralgia pure and simple. And so, again, of the so-called clavus there is every variety, from a form bordering closely on the migraine type to another, differing in nothing from an unusually severe ocular and frontal neuralgia of the fifth, except in the presence of a tremendously painful parietal focus. But the fact on which I would most particularly insist is one that was first taught me by my personal experience, viz., that migraine is, with extraordinary frequency, the primary or youthful type of a neuralgia which, in later years, entirely loses the special characters of scik-headache, and assumes those of ordinary frontal neuralgia, with or without complications. In my own case, the "sick-headache" character of the affection was strongly marked during the first two or three years, after which time it gradually but steadily lost all tendencies to stomach complications, and, what is more, the type of the recurrence became entirely changed. Yet it is quite impossible to believe that the malady is now a different one, in any essential pathological point, from what it was at first; if any disproof of this were needed, it might be remarked that the singular series of secondary trophic changes which have complicated my case have been impartially distributed between the respective periods when the affection was frankly migraineuse, when it was mixed, and when it was simply ophthalmic neuralgia (as it is at present ;) indeed, some of the most decided of these trophic complications (orbital periostitis, corneal ulceration, fibrous obstruction of the nasal duct) occurred within the period in which every attack of pain, unless I succeeded in getting to sleep very shortly, ended in violent vomiting. The experience thus gained has made me very attentive to the past history of those who, in later life, complain of frontal neuralgia without stomach complication, and it is surprising to find in how many cases patients, who at first declare that they never had neuralgia before, on reflection will recall the fact that they were often "bilious" in their youth; which "biliousness" turns out to have been regularly preceded by one-sided headache, and to have been severe in proportion to the severity and duration of that previous headache.

I ask the reader to dwell with fixed attention on this fact of the exclusiveness, or almost exclusiveness, with which the neuralgias of the anterior part of the head are represented during the period of bodily development, and especially in the years just succeeding puberty, by migraine or by clavus. When this fact has thoroughly entered the mind, we can hardly help joining with it that other and most important fact already noticed, of the close connection between the predisposition to migraine and the predisposition to epilepsy, and reflecting further on the strong tendency which epilepsy likewise shows to infest the earlier years of sexual life. In view of these things, it is difficult to avoid the inference that both the epileptic and the neuralgic affections of this critical period of life are the expression of a morbid condition of the medulla oblongata, in which the sensory root of the trigeminus has its origin; and further, that this morbid condition (tending to explosive and atactic manifestations of nerve-force) must have its basis in defective nutrition. For, be it remembered, the epoch of sexual development is one in which an enormous addition is being made to the expenditure of vital energy; besides the continuous processes of the growth of the tissues and organs generally, the sexual apparatus, with its nervous supply, is making by its development heavy demands upon the nutritive powers of the organism; and, it is scarcely possible but that portions of the nervous centres, not directly connected with it, should proportionally suffer in their nutrition, probably through defective blood-supply. When we add to this the abnormal strain that is being put on the brain, in many cases, by a forcing plan of mental education, we shall perceive a source not merely of exhaustive expenditure of nervous power, but of secondary irritation of centres like the medulla oblongata, that are probably already somewhat lowered in power of vital resistance, and proportionably irritable. Let us suppose, then, that to all these unfavorable conditions there was added the circumstance that the structure of the medulla oblongata, or of parts of it, was congenitally weak and imperfect; then surely it would be scarcely possible for these loci minimæ resistentiæ to escape being thrown into that state of weak and disorderly commotion which eminently favors pain in the sensory, and convulsion in the motor apparatus.

2. We have so far been mainly considering the relations to the production of neuralgia of certain conditions of the central nervous system which indisputably are inherent from birth. Let us now pass quite to the other extreme, and consider a class of momenta which take a decided part in producing many neuralgiæ, but which are altogether accidental and factitious, and cannot be included among the necessary hostile conditions of life. To push the contrast to the utmost, let us inquire first, what amount of influence in the production of neuralgia can be given by such a purely "functional" influence as educational misdirection of intellect and emotion?

It is somewhat strange, though every one accepts as a mere truism the maxim that sudden emotional shock may produce almost any degree or variety of nervous disorder, the slower but far surer influence of long-continued mental habit is often practically ignored. It cannot, indeed, be left out of sight as a cause of disorders of the mind itself, nor are there many who would deny that such diseases as cerebral softening are, in a considerable number of cases, the premature ending to a life that has been broken down by harrassing work and anxiety. But what is far less appreciated is the tendency of certain unfortunate mental surroundings and modes of mental life to produce a generally neurotic condition, which may express itself in a variety of functional disorders, among which not the least common is neuralgia.

I may fairly hope to be acquitted of any predisposition to lay exaggerated stress on this kind of influence in the production of neuralgia, considering all that I have said of the importance of that inevitable cause, the neurotic inheritance, and all that I shall have to say presently as to the effects of a variety of external influences of a totally different kind. But I confess that, with me, the result of close attention given to the pathology of neuralgia has been the ever-growing conviction that, next to the influence of neurotic inheritance, there is no such frequently powerful factor in the construction of the neuralgic habit as mental warp of a certain kind, the product of an unwise education. This work is not intended as a treatise either on religion or psychology, and yet it is impossible for me to avoid some few words that may seem to trench on the province of each : for I believe that there are certain emotional and spiritual and intellectual grooves into which it is only too easy to direct the minds of young children, and which conduct them too often to a condition of general nervous weakness, and not unfrequently to the special miseries of neuralgia. As regards the working of the intellect, it is easier to speak in a free and unembarrassed manner than respecting the other matters. There can be no doubt that, of intellectual work, that sort which exhausts and harrasses the nervous system is the forced, the premature, and the unreal kind; and this it is which predisposes, among other nervous maladies, to neuralgia. It is more difficult to speak the truth about emotional influences generally, and especially about those which are concerned with the highest spiritual matters ; but I should do wrong were I to suppress the statement of my convictions on this point. I believe that a most unfortunate, a positively poisonous influence upon the nervous system, especially in youth, is the direct result of efforts, dictated often by the highest motives, to train the emotions and aspirations to a high ideal, especially to a high religious ideal. It is not the object that is bad, but the machinery by which it is sought to be attained.

In modern society there are two principal methods which are popularly employed for this purpose ; I shall describe them, by two epithets which are selected with no offensive intention, as the Conventual and the Puritan methods of spiritual training. By the former is meant that kind of education which deliberately dwarfs the nervous energy, with the hope of preserving the mind from the contamination of unbelief and of sinful passion. It is a system which is not peculiar to the Roman Church, nor even to the Christian religion, and it need the less detain our attention, as its effects, so far as they are evil, are mainly seen in general nervous and mental enfeeblement, rather than in the outbreak of explosive nervous disorders, such as convulsion, insanity, or neuralgia. There are doubtless exceptions to the rule ; but that is the rule. It is far otherwise with the spiritual education which is here called Puritan, but which is confined to no party in the Church. This is a system which seeks to purify and exalt the mind, not by enforcing obedience to a series of spiritual rules for which another mind is responsible, but by compelling it to a perpetual introspection directed to the object of discovering whether it comes up to a self-erected spiritual standard. The reader will understand that I have not the remotest intention to depreciate either a true and manly self-restraint in obedience to the direction of "pastors and masters," or an honest watchfulness over one's own conduct and thoughts. But the lessons which our psychologists are rapidly learning, as to the evil effects on the brain of an education that promotes self-consciousness, are sorely needed to be applied to the pathology of nervous diseases generally, and of neuralgia among the rest. Common sense and common humanity, when united with the physician's knowledge, cry out against the system under which religious parents and teachers subject the feeble and highly mobile nervous systems of the young to the tremendous strain of spiritual self-questioning upon the most momentous topics. More especially is such a practice to be condemned in the case of boys and girls who are passing through the terrible ordeal of sexual development—an epoch which, as we have already seen, is peculiarly favorable to the formation of the neurotic habit, and I must emphatically state my belief that among the seriously-minded English middle classes, more especially, whose life is necessarily colorless and monotonous, the mischief thus worked is both grave and widely spread.

Perhaps the maximum of damage that can be inflicted through the mind upon the sensory nervous centres is effected when to the kind of self-consciousness that is generated by an excessive spiritual introspection there is added the incessant toil of a life spent in sedentary brainwork, and checkered with many anxieties, and many griefs which strike through the affections. Doubtless, such a combination of morbid mental influences is sufficient of itself to generate the neuralgic disposition in its severest forms, without any hereditary neurotic influence, and without any other peripheral irritations ; I have more than one such instance in my mind at this moment. But, if they can do this, much more can such influences arouse inherent tendencies to neuralgia ; to persons who are predisposed in this manner they are most highly deleterious.

3. We come now to the peripheral influences which in a more obvious manner become factors in the production of neuralgia. Of such influences there are an immense variety, and the only common quality that can be predicated of all is the tendency directly to depress the life of the sentient centre upon which their action impinges.

If we search among the external influences which contribute to the production of neuralgia for one that is apparently trivial as to the amount of material disturbance which it can cause, and yet is very frequently effective, we may select the agency of cold. The effect of a continuous cold draught of air impinging on the naked skin for some time is comparatively frequently seen in the provocation of neuralgic attack : we say comparatively, because this influence is more frequently effective than blows, wounds, or temporary irritations of any kind, applied to the peripheral ends of sensory nerves. But if neuralgia be a more frequent consequence of cold than of these other influences, a moment's reflection will show that it is by no means an absolutely common result. One has only to think of the numerous omnibus-drivers, engine-drivers, cab-drivers, etc., etc., who pass their whole working lives in presenting the (more or less) naked expanse of their trigeminal and their cervico-occipital nerves to every variety of wind, to to perceive that, were this sort of influence very potent in itself, male neuralgic patients should swarm as thick as bees in our hospital and dispensary out-patient rooms; which is notoriously quite contrary to the fact. The same remarks, in both directions, may be applied to the direct influence of atmospheric moisture, either with or without the effect of wind (of course I am not speaking of the more recondite effects of damp soil on the persons who live about it. [Among the hundred patients who formed the basis of the inquiries mentioned in this work, forty-one accused external cold of producing the attack, but many of these produced insufficient evidence that such was the case.] In short, the direct effects of atmospheric cold would seem to be these. Mere lowness of temperature goes for something, but not much; [The most marked instance of the effect of cold, *per se*, that I have seen, was exhibited by a young lady who was under my care during the past severe winter (1870–'71). During much of the time she was confined to a carefully-warmed apartment, on penalty of a violent paroxysm if she left it.] for about as much, perhaps, as it does in the way of aggravating all neurotic tendencies. Cold joined with wind is much more powerful. And the maximum of ill-effect seems reached by very cold wind mingled with sleet or driving rain, which keeps the skin soddon. But the conclusion at which I long ago arrived is, that none of these influences ever take more than a small (though it is sometimes an important) part in the production of neuralgia; and that in the majority of cases there is no pretence for supposing that they had the slightest share in its causation.

A word or two must be said as to the *modus operandi* of cold and cold wind, as these are the most frequent of external, so-called "exciting" causes. The popular use of such phrases as the latter has an extraordinary influence in disguising the plain fact, which is, that these influences operate wholly in the direction of robbing the nerves of force. The continuous abstraction of heat from the surface, which of course is materially aided by rapid movement of the air, must necessitate a readjustment of the distribution of energy, the only result of which must be to drain the sensory nervous centre of its reserve of force. But, in fact, there is an experiment, ready performed to our hands, which may amply satisfy us as to the kind of influence exerted by cold on superficial nerves, viz., the sensations experienced in recovering from frost-bite, which has been severe enough to paralyze the nerves without causing actual gangrene of the tissues. The passage of the nerves back from temporary death to full functional life is marked by a half-way stage in which there is agonizing pain.

4. We must next consider the effects of a class of peripheral influences which act, where they exist, in a more constant manner than any others; viz., those in which the trunk or periphery of a sensory nerve either receives a severe injury, or becomes more or less engaged in inflammatory processes, or compressed or otherwise damaged by the growth of tumors or the spread of destructive ulcerations.

With regard to ordinary nerve-wounds as a cause of neuralgia, we have already said (*vide* Chapter II.) nearly as much as it is necessary to say; we need only here point out that, like the influence of cold applied to superficial nerves, that of wounds must necessarily be a depressing one to the centre with which the wounded nerve is connected, and the resulting neuralgia must be regarded as an expression of impeded and imperfect nerve-energy, not of heightened nerve-function. The pain is set up during the process of nerve-healing; that is to say, at a stage intermediate between those of abolished function and completely restored function; and there can be little doubt that the obstinacy with which it is often protracted is due to the slowness with which a wounded nerve recovers its full functional activity; when once the latter is completely restored there is an end of neuralgic pain. It is exactly analogous to the course of events in recovery from freezing.

There remain for consideration, however, (a) a small class of cases of nerve-wounds in which the healing process is not simple; but the lesion is followed by the development of a tumor of the kind denominated true neuroma. The process consists of hyperplastic changes in the nerve-fibres; its commonest examples are seen in the extraordinarily painful swellings that occur on the ends of nerves left in stumps after amputations; but, in fact, a neuroma of this kind may occur after any kind of severe nerve-injury, as, *e. g.*, a cut from broken glass, the impaction of foreign bodies, etc. The true neuromata are composed mainly of nerve-tissue, with a relatively small element of connective tissue: the nerve-fibres can be traced directly to the nerve-tumor. Besides the traumatic neuromata which form permanent tumors, incapable of being got rid of except by actual excision, a minor variety of the same kind of change has in several cases been known to take place in consequence of an abiding local irritation from the impaction of a foreign body, on the removal of which the neuromatoid enlargement completely disappeared. (b) There are likewise a certain number of cases in which a tumor is developed from the neurilemma, and does not consist of nervous tissue; these are distinguished as false neuromata, and may be of various kinds, the fibromatous and gliomatous being far the most common, but cysts and cystic tumors also sometimes occurring.

The case of the neuromata is well worth reflecting upon, in the course of our endeavors to clear up the Pathology and Etiology of Neuralgia. If ever we could find a merely peripheral influence which would of itself be invariably competent to excite neuralgic pains, it would surely be found in neuroma; but the case is not merely not so, it is strikingly contrary. Just as wounded and inflamed nerves frequently go through the whole processes of disease and recovery without once eliciting a neuralgic pang, so is it with neuromata; they are not unfrequently quite indolent, and neither excite neuralgia, nor are themselves at all particularly tender to the touch. And what is most remarkable is, that, as Eulenburg correctly remarks, among the pseudo-neuromata the kind of tumor which is most frequently associated with neuralgia is by no means the dense fibroma or glioma, which might be expected by its mechanical pressure to excite inevitable neuralgic pain, but the far softer and more yielding cystic tumors. I do not know how the facts may affect the reader, but to me they suggest the strongest possible arguments against the belief that peripheral irritation can of itself produce neuralgia without the intervention of some centric change. The tendency to such change (from inherent constitution) in the sensory root of the nerve must surely be the reason why neuroma causes neuralgia in a given number of subjects, instead of letting them go scot-free, as it does other persons.

The same remarks apply to the result of observations on the effect of tumors commencing in tissues altogether unconnected with the nerve, and merely coming to involve it, secondarily, in pressure. It has been often noted that, among these tumors, fluid-containing cysts and soft medullary cancers are far more frequently the cause of decided and distressing neuralgia than the denser and less yielding neoplasms. Of kinds of tumors that are specially apt to produce severe and even intolerable neuralgia by the pressure on nerves, it has been remarked that aneurisms are among the worst: here every pulsation often sends a dart of agony through the nerve. There is a reason here, however, which is often left out of sight; not merely is the perpetually varying pressure specially harassing and exhausting to the nerve, but in many of these cases there is general arterial degeneration, and the sensory root of the nerve is exceedingly likely to be very badly nourished. [This result will be more directly brought about when the aneurism happens to press on the ganglion of a posterior root.] We pass now to the consideration of the influence exerted by other great series of peripheral impressions in the production of neuralgia. These impressions are connected chiefly with the functions of the digestive and of the genito-urinary organs, the functions of the eye, and the nutrition of the teeth.

To take the least important of these first, I may surprise some readers by the statement, which I nevertheless make with much confidence, that irritation of any part of the alimentary canal is, on the whole, a rare concurrent cause, even in the production of neuralgia. There are, as has been already fully explained, cases of neuralgia seated in these viscera themselves (or the plexuses in their immediate neighbourhood), although their number is immensely smaller than that of the neuralgias of superficial nerves. But it is not at all common—it is even exceedingly rare—for irritation conveyed from the alimentary canal to take any important part in setting up neuralgia of a distant nerve, even when that nerve has close connections, through the centres, with those coming from the irritated portion of the alimentary canal. Valleix had the great merit to perceive this, even in the case of neuralgias of the head, where appearances are so likely to lead the observer to a contrary opinion. And it is not a little remarkable that this should be the case, when we consider the close central connections which the vagus, the great sensory nerve of a large portion of the alimentary canal, has with the sensory root of the trigeminus. In fact, however, there are certain peculiar forms of gastric irritation which do react upon the trigeminus; for instance, a lump of unmelted ice, suddenly swallowed, almost invariably produces acute pain in the supra-orbital branch of the fifth, on one side or the other, and occasionally (as in a case cited by Sir Thomas Watson) in other nerves, But that common dyspeptic troubles at all frequently or importantly contribute to the production of neuralgia, I do not for a moment believe: it needs some very powerful irritation, such as that just mentioned, or as impaction of great masses of scybalæ in the intestines, or severe irritation from worms, to produce such an effect.

It is far otherwise with the genito-urinary apparatus; in a large number of cases, irritations proceeding from these organs do undoubtedly contribute to the production of neuralgia, though by no means in the important degree which many authors seem to have assumed. There can be no doubt, for example, that the irritation of a calculus, either within the kidney itself, in the ureter, or in the bladder, may set up violent neuralgia, which for the most part is localized in the branches of the lumbo-abdominal nerves. The instance of the eloquent Robert Hall is an example of renal calculous acting in this way: he suffered the most excruciating agony for years, and was obliged to take enormous quantities of opium in order to make life endurable. An instance of calculus impacted in the ureter, in a gentleman somewhat past middle age, occurred in my own practice; the lumbo-abdominal neuralgia occurred in frequent paroxysms of dreadful severity; and another case, already referred to was that of a woman, in whom ovarian neuralgia was undoubtedly in great part due to the irritation of an impacted calculus in the ureter. These cases, however, are very rare in comparison with others in which the peripheral source of the neuralgia is either the uterus or ovary, or the external genitals. I have no means of ascertaining, with anything like accuracy, the frequency with which the internal sexual organs are the starting-point of neuralgia, because the majority of such cases pass, naturally, to the care of physicians who practice chiefly in the diseases of women, and consequently not adequately represented either in my hospital or my private practice; still, I have seen a good many of these affections, and, though I speak with the reserve necessitated by the circumstances just named, I am much inclined to believe that even such powerful centripetal influences as those of the states of commencing puberty, of pregnancy, of the change of life, and uterine diseases generally, are very rarely the cause of true unilateral neuralgia, except in subjects with congenital tendencies to neuralgia. But in predisposed subjects there can be no doubt that these influences assist most powerfully in producing the malady.

Of the power of irritation of the external genitalia to act as a so-called "exciting cause" of neuralgia, there is abundant evidence. I would especially call attention to the remarkable monograph of M. Mauriac, ["*Etude sur les Nevralgies Reflexes symptomatiques de l'Orchi-epididymite blenorrhagique*" Par C. Mauriac, Medecin de l'Hospital du Midi. Paris, 1870.] on the neuralgias consecutive to blenorrhagic orchi-epididymitis, as illustrating this with a force that was to me, for one, surprising. I shall, perhaps, have further occasion to these researches; here it will be enough to mention that M. Mauriac's enormous experience of blenorrhœa and orchitis at the Midi has shown that, in an exceedingly large number of cases, certainly not less than four per cent., this combination is followed by reflex neuralgias, of which a large number are not seated in the genital apparatus, but affect the track of some distant sensory nerve, though the intermediation of the spinal centres; and that with these reflex pains there is often profound general disturbance, including very often an extremely profound general anæmia. The most frequent kind of these neuralgias is rachialgia, *i. e.*, pain in the superficial posterior branches of spinal nerves; next comes lumbo-abdominal neuralgia; then sciatic and crural, visceralgic (abdominal), etc.; and besides all these there are numerous instances of neuralgia in the testis. As to the nervous "reflection," more hereafter.

It has surprised me, somewhat, that while M. Mauriac has seen so many reflex neuralgias set up by orchi-epididymitis, he does not appear to have noticed cases of trigeminal neuralgia from this source; because, in the very analogous instance of the peripheral irritation produced by excessive masturbation, we undoubtedly do frequently get a development of the tendency to migraine, and also to other forms of neuralgia of the fifth: moreover the effect of such local irritation can be occasionally traced with much distinctness in the trigemini, by a tendency to certain forms of eye-disease without positive neuralgia. This was remarkably exemplified in a case which was under my care some years ago, and in which both eyes were greatly damaged by vaso-motor and trophic changes; partial insanity also supervened with hallucinations of sight and hearing.

We come now to one of the most powerful sources of peripheral irritation tending to set up neuralgia; viz., functional abuse of the eye. This is one of the very few peripheral influences which occasionally we see producing neuralgia unaided by hereditary predisposition, or any other observable cause whatever, and in a far larger number producing it with the sole aid of more or less defective general nutrition. The latter occurrence is well exemplified by a case which Mr. Carter sent me the other day, and which also illustrates (second attack) the effect of the superaddition of syphilitic taint:

Matilda W——, aged thirty-three, married, and has three very healthy children. Comes of a remarkably healthy family, of which she told me the entire history for three generations, with unusual intelligence and clearness. No neuroses, properly so-called, in any of her relatives during all this time. She herself was a very strong and hearty girl until the age of seventeen; between this date and her marriage, three years later, she was obliged to work tremendously hard at fine sewing, by which means she gained a very scanty livelihood. After a comparatively short period of this work she began to suffer from typical attacks of migraine, very severe, and recurring every three or four weeks, but in no particular connection with the menstrual function, which was normal. On her marrying and ceasing to do needle-work, the migraine entirely disappeared, and she retained perfect health till the commencement of 1871. At this time she had suckled a very hearty baby for ten months, and was not able to furnish such good living as usual. She was attacked early in January, with violent neuralgia affecting all three branches of the right fifth, and she the more readily applied for advice because she soon found that the neuralgia was becoming complicated with dimness of vision in the eye of the affected side, "as if she was going to have a cast." Was quite unconscious of ever having had syphilis. The medical man encouraged to believe that the whole malady was nervous, and would soon disappear under appropriate remedies, and gave her quinine, under which treatment she declares that she was rapidly improving, both as to pain and vision, but that her resources came to an end, and she could no longer pay for the medicine. She then neglected herself, and rapidly got worse in all regards, till at last she was compelled to apply to the South London Ophthalmic Hospital, whence Mr. Carter sent her to me, on the 6th of April. At this time the paroxysms were excessively violent and frequent, though brief. On examination, tender points were found at the supra-orbital notch, at the infra-orbital foramen; in front of the ear; in the temporal region; in the parietal region, and the inferior dental region. There was strongly marked anæsthesia of the skin of the right half of the face, of the gums, and of the side of the tongue. The teeth were absolutely perfect: not one spot of caries could be seen. Taste was completely destroyed in left half of anterior part of the tongue. Smell was totally lost on both sides, and had been so, the woman declared, from a very early period in the illness. The right eye showed complete paralysis of the levator palpebræ and of the external rectus; nearly complete paralysis of the superior and inferior rectus, rather less marked paralysis of the internal rectus. Pupil normal, conjunctiva moderately congested, lachrymation profuse, photophobia partial. The functions of the retina were perfect. Accommodation was affected in the following degree and manner. The vision of the affected eye was perfect at long distances, very imperfect at short distances. With both eyes open she saw every thing double, but could still count all the bricks in a whitewashed wall· at sixteen feet distant. There was no secondary disturbance of the stomach whatever. On the first visit she assuredly had no visible signs, in skin or throat, of syphilis; the perfect health of her children, and absence of abortions, made syphilis the less probable. But on her second visit she complained of sore throat, and a week later a palpably specific sore appeared on the soft palate. She declared, with apparent sincerity, that it was the first symptom of the kind she had ever had. The neuralgia rapidly disappeared under thirty grains of iodide of potassium daily. The lesions of taste and smell disappeared exactly pari passua with the trigeminal pains. The ocular paralysis threaten to be much slower in departing. I think we must believed that this woman contracted syphilis after the birth of her last child. It is at any rate certain that the migraine of her youth was perfectly unconnected with syphilis, being as unlike the pains evoked by the latter as it is possible for two kinds of pain to be. In all probability she was infected during her last lactation.

Last among the peripheral influences of sufficient importance to be specially mentioned as effective factors in the production of neuralgia, must be mentioned caries of the teeth, and the comparatively rare accident of the mal-position or abnormal growth of a "wisdom-tooth." It is an undoubted fact that these things may cause neuralgia even of a very serious type, and attended with extensive complications ; as in Mr. Salter's cases, already mentioned, of reflex cervico-brachial neuralgia from carious teeth. Looking to the extreme frequency of caries, however, as compared with the rarity of true neuralgia (not mere toothache) as a consequence of it, it is impossible not to suppose that the share of the carious teeth in the production of such neuralgia must be very small, compared with that of other influences.

5. The next influence which we shall mention as undoubtedly very effective in assisting the production of neuralgia in certain cases is that of anæmia and mal-nutrition generally; but it is not necessary to dwell on this at any length. The fact is notorious that severe loss of blood is always followed by headache; and if there be the least predisposition to neuralgia, this headache will very commonly take the form of the severest clavus. And, in like manner, chronic states of anæmia and of mal-nutrition undoubtedly aggravate every existing neuralgia, and bring out lurking tendencies to the disease. But I do not believe that anæmia, or starvation pure and simple, ever generates true neuralgia by its sole influence.

6. The question how far, and in what way, the neuralgic tendency is helped by certain constitutional diatheses, such as rheumatism and gout, and by certain toxæmiæ, such as malaria alcoholism, lead-poisoning, etc., is a very much more difficult one than might be supposed from the off-hand manner in which many writers speak of the "rheumatic," the "gouty," or the "alcoholic" forms of "neuralgia." We may, however,

simplify it a good deal. In the first place, it seems obvious to me that the only manner in which alcohol helps the production of true neuralgia is by its tendency, after long abuse, to produce degeneration of the nervous centres: it will therefore be considered, shortly, under another division of the present subject. Lead-poisioning, again, only produces so highly special a form of neuralgia (if colic be neuralgia at all) that it need not detain us here. The influence of malaria is, for the most part, an utter mystery to us, but by so much as we can see it appears plain that one of the most important features in the disease is a powerful disturbance of the spinal vaso-motor centres. But the most interesting consideration that we have to deal with is the question of the supposed relations of the rheumatic and the gouty diatheses, and the syphilitic dyscrasia, to the neuralgic tendency. On this point I am obliged to disagree *in toto* with the popular view that assigns these diatheses among the most frequent predisposing causes of neuralgia.

To take the case of rheumatism first, I am willing to allow that there are a number of facts which superficially appear to countenance the idea of a close connection of this disease with neuralgia. But of these facts a considerable proportion consist only of examples of inflammation of the nerve-sheath, with a certain amount of effusion within and around it, occurring in persons who have never shown any symptoms which warrant the assumption of a general rheumatic diathesis ; and these local phenomena really differ in nothing from many trophic and vaso-motor changes which have been already described as plainly secondary to ordinary neuralgia in which there could be no pretence of a rheumatic pathology except on the slender foundation of a suspicion that the affection was immediately excited by the influence of cold, which is really no argument at all. Such patients will be found to have exhibited, not special rheumatic, but special neuralgic tendencies in their past history. On the other hand, there undoubtedly are a certain number of patients who, having previously given signs of a tendency to generalized rheumatic inflammation of fibrous membranes, are, on some particular occasion, attacked with similar inflammation extending over a more or less considerable tract (not a small limited spot) of a nerve sheath. But so far from agreeing with those who think that this is a frequent case, my experience teaches me that it is quite exceptional; nor do I believe that the common opinion could ever have arisen had it not been for the rage that exists for connecting every disease with a special diathesis which the profession flatters itself that it understands. Few persons have taken more pains than myself to ascertain the frequency with which neuralgic patients show a history of previous rheumatism, whether in the so-called "fibrous," or in the synovial form; but it is remarkable how seldom I have found this to be the case—a result which surprised me, because it happened that I, a neuralgic subject, had suffered in youth from regular acute rheumatism, and had fancied that I should discover a close connection between rheumatism and neuralgia. Eulenburg states that neuralgia caused by cold more frequently attacks the sciatic nerve than any other, and thinks that the tendency to sciatica is characteristic of the relations of rheumatism to sensory nerves. For my own part, I see no reason to call in the rheumatic diathesis as a *deus ex machina* to explain the frequency with which sciatica follows comparatively trifling peripheral impressions like that of cold. The true reason I believe to be, that what would have been a slight and trivial neuralgia elsewhere, becomes a serious affection in the instance of the sciatic nerve, by reason of the strong muscular pressure end dragging which are always going on in the thigh in locomotion. I shall return to this subject when speaking of Treatment.

As regards the relations, of gout to neuralgia, I can hardly express my own view better than by quoting the words of Eulenburg:* "Much more doubtful is the influence of gout, which in rare cases, perhaps, produces neuralgia directly, by means of neuritis, or by the deposit of tophus-like calcareous concretions in the nerve-trunks. Gout has been reckoned as a great influence among the causes of superficial neuralgias (sciatica), and also of visceral neuralgia (angina pectoris, etc.,) but this influence is more probably only an indirect one, operating through circulation changes which are often produced by chronic liver-diseases or by diseases of the heart and vessels, (*e. g.* Valvular diseases and narrowing of the coronary arteries in angina)." To which I will add this argument against any close connection of gout with neuralgia, that it is exceedingly seldom that colchicum effects any decided good, a fact which is as unlike the relations of colchicum to true gout as any thing could be. For, whatever may be thought of the advantages or disadvantages, on the whole, of employing colchicum against gout, at least no one with any experience will deny that in the immense majority of cases of true gouty pain, it gives rapid relief to the acute suffering. I doubt if it ever † acts in that way in real neuralgia. though I have occasionally seen it apparently useful in a more limited way, as will be said hereafter.

As regards the relation of the syphilitic dyscrasia to neuralgia, I agree in general with Eulenburg. "Syphilis," he says, "may be the direct cause of neuralgia, either by the develop-

* *Op. cit.*, p. 60.

† This opinion is somewhat stronger than that expressed in my article in the "System of Medicine." I can only say it is the result of much increased experience.

ment of specific gummata in the nerve-trunks or in the centres, or by arousing chronic irritative processes in the nerve sheaths, the membranes of the brain and spinal cord, or, especially, in the bones and periosteum (syphilitic osteitis and periostitis).' The case of periostitis, however, is a doubtful one: it may be questioned whether this affection (which will be among the diseases discussed in Part II. of this work) ever give rise to true neuralgia. Persons who are, by inheritance, highly predisposed to neuralgia, may from the mere general lowering of their health produced by constitutional syphilis, become truly neuralgic simultaneously with, or subsequently to, the appearance of painful nodes on their bones. And as regards the whole relations of syphilis to neuralgia, I must, from my experience, conclude that the former is, after all, but rarely concerned in the production of the latter. Syphilis has a strong specialty for producing limited motor paralyses, but a much weaker one for producing limited affections of the sensory system.

7. We now come to the discussion of a group of momenta whose influence in the production of neuralgia is at once very powerful, and of the highest significance as regards the general pathology of the disease. These are the degenerative changes of the arterial and capillary systems which are a part of the normal phenomena of old age, but may occur at earlier periods of life, in consequence either of certain constitutional diseases, especially gout, or of special toxic influences on nutrition, of which persistent alcoholic excess is very far the most important.

The reader does not need to be told the familiar story of the degenerative changes in the vessels which, commencing usually some time during the fifth decenniad, by degrees convert the elastic arterial coats, and the almost membranous walls of the capillaries, into more or less rigid tubes; nor does he need to be informed that the tendency of these changes, as they operate in the great motor and intellectual centres, is notoriously to produce innutrition of the tissues that depend for their blood supply on the affected vessels, whence cerebral softening so commonly results. That analogous changes take place in the vessels supplying the spinal centres is certain; but it is a remarkable fact that these do not very commonly produce motor paralysis. What they do produce is rather a slow enfeeblement both of (spinal) sensation and motion, but where the process of decay has been prematurely forced, or the inheritance of neurotic weakness is very marked, the process of sensorial decay (the decline, that is, of true sensorial function) is apt to be mingled with pain. That this pain should be localized, often in a single nerve, is no more surprising than the fact that the degenerative process itself should vary so greatly in the degree of its development at one point from that which it shows at others. I have already insisted (*vide* Chapter I.) on the marked correspondence between the period of life in which degenerative changes commence and progress (the last third, roughly speaking, of a fairly long life), and that in which the most severe, intractable, and progressively increasing neuralgias are developed. I must here notice a singular statement of Eulenburg's, that neuralgia never attacks people who are over seventy. That statement shows that persons of a greater age than seventy are rare in this world, and that no such patient happened to come under Eulenburg's notice; for I have (by mere chance, doubtless) seen several instances of first attacks occurring after seventy; and almost the worst case of epileptiform tic I ever saw began when the patient was eighty; she was a member of a highly neurotic family whose medical genealogy is given at a previous page. In general terms, it may be said that every additional year of life after fifty increases the probability that a neuralgia, should such arise, will be severe and rebellious to treatment; and in the very aged the cure of such affections is probably impossible.

8. This seems the proper place to introduce such facts as have been observed, and they are very few, that directly illustrate the material changes occurring in neuralgia.

Very much the most important of these facts is the history of a remarkable case recorded by Romberg. ["Diseases of Nervous System," Syd. Soc. Trans., vol. i.] The patient, a man sixty-five years old at the time of his death, had suffered for several years from the most violent and intractable epileptiform trigeminal neuralgia, complicated with interesting trophic changes of the tissues. Post-mortem examination showed that the pressure of an internal carotid aneurism had almost destroyed the Gasserian ganglion of the painful nerve, that the trunk and posterior root of the nerve were in a state of advanced atrophic softening, and the atrophic process had extended in less degree to the nerve of the opposite side. Now, the value of this case is by no means restricted to the fact that it records the existence of a particular anatomical change in one example of neuralgia. Its most striking teaching is the fact that the acutest agonies of neuralgia can be felt in a nerve, the central end of which is reduced to such a pitch of degeneration that conduction between centre and periphery must very shortly have entirely ceased had the patient lived. And hardly less important is its illustration of the fact that permanent injury to the ganglion of the posterior root of a spinal nerve impairs the vitality of the posterior root itself—a fact which has been independently made out by the physiological researches of Bernard and of Augustus Waller.

On the other hand, if we examine the tolerably numerous histories of cases in which the painful nerves have been examined at the apparent site of pain, we discover nothing to lead us to connect neuralgia definitely with any one sort of change. Assuredly, for example, local neuritis is by no means universally, it is probably even not commonly, present in the early stages of neuralgia; it has also been repeatedly detected in nerves that had been wholly free from neuralgia; and, on the other hand, it has been entirely absent in nerves that have been the seat of the severest pains. Moreover, many facts which have been put down without reflection, as showing a local peripheral cause for neuralgia, are at least open to another and, as I believe, truer explanation; as (e. g.) in the following remarks of Eulenburg on mechanical irritations of nerves as causes of neuralgia: "Diseases of bones are extraordinarily frequently the cause of neuralgias in consequence of compression or secondary disease, which affects the branches of nerves passing through canals, foramina, fissures, or over processes of bone. The appearances which the opportunities of resections of the trigeminus for facial neuralgia have permitted to be discovered, have given us valuable information in that direction. Flattening and atrophy of nerves from periostitis, or from concentric hypertrophy in narrowed bony canals, have frequently been discovered. The neurilemma at the narrowed parts was often seen reddened, ecchymosed, infiltrated with serum, or surrounded with fibrous exudation; occasionally inflammation had been followed by partial thickening of the neurilemma (fibrous knots) and turbidity (Trubungen) of the nervous cord at the corresponding spot. Similar appearances have been noted in other neuralgias (neuralgia-brachialis, sciatica)." For my own part, I believe that the above description represents the facts from an erroneous point of view. True neuralgia, if by that we understand a pain of intermittent character limited to one or more nerves, is in my experience an extremely uncommon result of periosteal disease, or of inflammation of the linings of bony canals; but in a great number of instances such diseases appear to be set up as the secondary consequence of the neuralgic process (whatever the essential nature of that may be) going on in sensory nerves which supply the parts when these inflammations appear. And it must be remembered that the specimens obtained by resection of nerves are comparatively few in number, and are taken universally from old-standing and desperate cases of disease; in short, from cases which are just in those advanced stages of neuralgia in which, as has already been amply shown, these secondary inflammations are almost always present. On the other hand, I have myself had one opportunity of examining the local condition of an intercostal nerve, which during life, and quite up to death, had been the site of the most pronounced neuralgia, which, however, had only existed for a few days. The patient, a young man, aged twenty-seven, was probably insane, and had attempted suicide. Not a trace of inflammation, either in the nerve itself or in any of the tissues to which it was distributed, could be detected. (This was a case in which I greatly regretted the impossibility of getting a family history that was at all reliable.) The spinal cord, unfortunately, could not be examined. And I strongly believe, from the marked absence of tenderness on pressure which is almost universally observed in ordinary cases of neuralgia at an early stage, that primary inflammation of neurilemma, periostem, etc., as a cause of neuralgia, is altogether exceptional; so much so, that we are entitled to believe it can never be more than a concurrent, and then not the most important, cause.

It is necessary here to inquire, more particularly than we have yet done, into the nature of the "painful points" first signalized by Valleix as a distinctive symptom of neuralgia. Very great differences of opinion have prevailed among subsequent writers, both as to the frequency and the significance of these points. It may be said, however, to be now quite settled that the presence of definite points, painful on pressure, and also corresponding to the foci of severest spontaneous pain, is far from universal in neuralgia. Upon this point there is probably no reason to doubt the correctness of Eulenburg's observations made in the surgical clinic of Greifswald and the polyclinic of the University of Berlin; he says that he discovered the existence of tender points in "Valleix's sense," in rather more than half the cases of superficial neuralgia, but in the rest he could not by any means discover them. In many other cases, however, he found more indefinite points of tenderness, not accurately corresponding to nerve-branches, but affecting individual portions of skin, bone, or joints; the relation of these to the neuralgic symptoms was difficult of explanation. Eulenburg lays down the principle that "hyperæsthesia" may depend on three sorts of causes—(1) On local disease of the peripheral ends of nerves; (2) on alterations of the psychical centres; and (3) on morbidly exaggerated conduction in the nerve-trunks themselves; and it is to this third source that he attributes many of the phenomena of the neuralgic painful points, and especially their multiplicity, in many cases. The *locus in quo* of the mischief which sets up this exaggerated conduction of sensory impression is, upon this theory, between the psychical centre and the main point of branching of the nerves; hence a large number of peripheral nerve-termini might be practically sensitive to touch, because the mischief, though localized in a comparatively small spot, might easily affect many bundles of fibres, which diverge widely from each other in their course. It will be seen presently with what limits and for what reasons we believe this to be a true theory. But to return to the question of painful points in Valleix's sense, we must state one or two facts which seem certain from our own experience, but have not been adequately recognized, we believe, by others. The first is, that localized tender spots, accurate pressure on which will set up or aggravate the neuralgic pain, are not early phenomena, save in neuralgias of exceptional severity of onset; but that a certain persistence and severity of neuralgia are always followed by the formation of one or more true points douloureux. The second fact relates to the clinical history of migraine. Roughly speaking, it is true, as Eulenburg states, that in pure migraine, painful points in Valleix's sense are not to be found; in place of them we observe, after the paroxysms have passed away, a more generalized soreness of considerable tracts of the scalp, forehead, etc., or diffuse tenderness of the eyeball. But I must here again refer to the fact, first observed in my own case, and afterward verified in many others, that migraine may be only the youthful prelude to a regular trigeminal neuralgia attended with the formation of characteristic localized painful points at a later period. And the third fact that must be specially mentioned is that the true Valleix's point, when it has become established for some time, is not a mere spot of sensitive nerve, but is the scene of trophic changes, involving hyperæmia and thickening of parts surrounding the nerve. To give one example, it is quite a frequent thing to find a patch of tender and sensibly thickened periosteum of irregular shape, but equal sometimes to a square inch in size, over the frontal bone at and immediately above the inner end of the eyebrow, in cases where supra-orbital neuralgia has recurred frequently during some years, although no such thing was present when the neuralgia first commenced. In my own case, the bone has become sensibly thickened at that point.

The general result of such post-mortem and clinical information as can be had seems clearly to be that positive anatomical changes, either of nerve-terminals or superficial nerve-branches, are but casual and infrequent factors in the first production of neuralgia, and, in particular, it would seem that inflammation of a nerve itself by no means necessarily produces neuralgic pain, but (far more commonly) simple paralgesia or anæsthesia of the parts external (peripheral) to the lesion. The one marked exception to this general proposition is to be found in the case of the severe and peculiar injuries inflicted on the trunks of nerves by gunshot-wounds which, as we have seen (from the American experiences), can produce some of the most dreadful forms of neuralgia. But the nature of the injury here inflicted is, it must be remembered, quite different from any thing which either disease or accident in civil life would produce, save in the most exceptional instances. For the chief material element in the production of the neuralgias of ordinary life we are really driven, by exclusion, to the condition of the posterior roots of special nerves, in some cases, perhaps, to the (spinal) ganglia on which the nutrition of these roots probably is considerably dependent.

With the field thus narrowed for us, it is surely legitimate, in the necessary scarcity of anatomical records referring directly to the state of the nerve-roots in ordinary neuralgia, to place great weight on the facts of a disease like locomotor ataxy, in which the main anatomical change is a progressive atrophy of the posterior columns which usually falls with peculiar severity on the posterior nerve-roots, or on the parts of the gray matter immediately adjoining these, and in which neuralgia may be said, for practical purposes, to be a constant and most characteristic phenomenon. If any one desires to see how strikingly the connection of the neuralgic phenomena with the anatomical-change comes out, I rocommend him to study Dr. Lockhart Clarke's papers on locomotor ataxy (*vide* "St. George's Hospital Reports,i' 1866; *Lancet*, June, 10 1865; "Med.-Chir. Soc. Transactions," 1869), or the excellently reported case by Nothnagel (*Berlin Klin. Wochensch.*, 1865). It is really not too much to say that the only important difference between the, clinical aspect of the pains of locomotor ataxy and those of ordinary neuralgia is simply such as depends on the fact that the anatomical change in the former case is bi-lateral, and usually affects the roots of several, sometimes of a great many pairs of nerves. I infer, from a conversation with Dr. Clarke, that he fully recognizes the force of the analogy, and the great strength of the presumption which it sets up in favor of an atrophic change of the posterior roots in neuralgia.

It may, of course be urged, against the view that neuralgia depends on any change analogous to those which occur in ataxy, that quantities of cases of the former recover speedily, and must be supposed to be either independent of material change altogether or, at any rate, to have involved only very trivial anatomical changes, not formidable diseases, like atrophy of nerve-centres. I find it impossible to admit that this argument has the slightest force. Are we to suppose that the posterior nerve-roots alone, of all tissues and organs of the body, are incapable of minute and partial changes in the direction of molecular death which may be perfectly recovered from in weeks, months, or even days? I, for one, cannot doubt, that such changes are of frequent occurrence, in all parts of the central nervous system, when I can consider the absolute dependence of these portions of the organism upon a perfect blood-supply, and the immense number of possible causes of temporary interference with that source of nutrition. And I can see no probable difference, except in degree and persistence between the effects on sensation which would be produced by such a change of the posterior roots as this, and that which would result from the more serious and fatally continuous change which is involved in locomotor ataxy.

9, We come now to a most important but most complex and difficult portion of the argument respecting the *locus in quo* of the essential pathological process (if such there be) in neuralgia; viz., as to the paths and the character of the so-called "reflex" influences which intervene in the causation, both of neuralgia itself, and also of the numerous complications with which we have seen that neuralgia is liable to be attended. The clinical facts which confront us here, and demand explanation, are the following: (1) Irritation so called, of sensory fibres may apparently evoke pains attributed to the site of the irritation, or to the parts on the peripheral side which are supplied by the same sensory nerves (2) Peripheral irritation of a particular sensory nerve may evoke neuralgic pains in nerves connected with that irritated only through the spinal centre. (3) Neuralgia in a sensory nerve may (and almost always does, to some extent) produce secondary vaso-motor paralyses: these paralyses may affect fibres which run in the same branch of the nerve as that which is painful, or fibres that run in another branch of the same nerve, or fibres that run with another sensory nerve, or the ganglionic chain of the sympathetic itself. (4) In like secondary manner, neuralgia may produce vaso-motor spasms in any of the directions just specified; this is usually a short-lived phenomenon, giving place quickly to paralysis; but Du Bois Reymond's often-quoted analysis* of his own sufferings from migraine seems to show that spasm-producing irritation of the trunk of the sympathetic may last during some hours. (5) Neuralgia in a sensory nerve may increase, alter, or (more rarely) suspend the secretions of glands supplied by fibres bound up either in the same branch, or in another branch of the same nerve, or in a different nerve with which it is connected only through the centre or (possibly) only through a plexus. (6) Neuralgia in a sensory nerve can produce paralysis of muscles supplied by motor fibres bound up with the painful branch, or with another branch of the same nerve, or in muscles supplied by a totally distinct nerve connected only through the centre. (7) It may produce convulsion and spasms of muscles, in all the above directions; this usually alternates with great weakness, or actual paralysis of the same muscles. (8) It may produce partial or complete loss of common or special sensation in nerve-fibres that run either with the same branch, or with another branch of the same nerve. (9) It may produce trophic changes, either in the direction of simple atrophy or of sub-acute inflammation

Journal de la Physiologie, v.

with proliferation of lowly-vitalized tissue (*e. g.,* connective) in the parts with which are supplied with sensation by the painful branches or by other branches of the same nerve.

It is necessary to go over again the proof of these facts; they are given pretty copiously in the chapter on Complications; and could have been made much more numerous. But the point to which I desire to compel the reader's attention is the impossibility as it seems of me, of accounting for the varieity and complexity of these phenomena, except by the supposition that there is in every case of neuralgia 'a central change, which is the one most important factor in the producing both of the pain and of the secondary phenomena. For the result of my experience is that neuralgia, unless very slight and brief, is never unattended by these complications and in the great majority of cases involves several different secondary alterations of function which must (so to speak) radiate from the central end of the sensory nerve, and from no other place whatever. And it must be remembered that the most elaborate "*symptome-complexe*" (is found equally in cases where no suggestion of any peripheral origin of the pain can be made, and in cases where, at first sight, one might fancy there was a very obvious peripheral cause for pain. I am quite willing to admit, with Eulenburg and others, that the evidence, powerful and varied though it be of the relations of neuralgia to hereditary neuroses, to alcoholic and senile degeneration, etc., only raises a strong probability that some part of the central nervous system is the *locus in quo* of the essential morbid processes in the majority of neuralgias. But the case stands far otherwise now that we are able to show, not merely that the majority of neuralgic patients suffer from such influences as those above mentioned, but that every variety of neuralgia is liable to be complicated with secondary affections of the most divergent nerves, the only common meeting-place of which is in the spinal centre of the painful nerve; and when we find moreover, that many of these secondary affections can equally be produced by undoubted atrophic changes (as in ataxy of those same posterior roots.

At this point we must introduce a remark relative to the true nature of so-called "reflex" effects. The word is constantly used, and is also much abused, as Eulenburg remarks. We all understand, of course, what is intended by the commonest use of the word : the case of sneezing produced by the irritation of snuff applied to the peripheral branches of the fifth nerve in the nose is a stock example. But another application of the phrase, of much more questionable propriety, is that where it is employed to designate functional nervous actions, which merely arise simultaneously with or subsequently to sensory phenomena as to which there is no proof whatever that they were produced by peripheral irritation. This particular inaccuracy of customary speech has probably contributed largely to the inveteracy with which writers on nervous disease have insisted on assuming a peripheral origin in every case for neuralgia itself. In the case of sciatica, for example, complicated, secondarily, with paralysis of the flexors of the limb, it seemed easy and scientific to speak both of the neuralgia and the paralysis as "reflex" effects of a local peripheral mischief—gouty, rheumatic, or the like; and it appears to have been perfectly forgotten by many that the whole phenomena might be explained by an original morbid action in the sensory root of the nerve, extending subsequently to the motor root, without any intervention of peripheral irritation whatever, or under the influence only of the ordinary peripheral impressions, which, in health, evoke no painful nor paralytic symptoms. It is by this kind of extension of a central morbific process, leading to radiation of the perturbing influence centrifugally along divers nervous paths, that I believe we must explain the facts observed in complicated cases.

Take, for example, the following case, which, in its history of twenty-three years, presents a fair example of a type of trigeminal neuralgia which I believe to be the rule rather than the exception, though the trophic changes were somewhat unusually varied and interesting. The following would be the pathological order of events, according to the radiation theory: First or true migrainous stage; failure of nutrition of a portion of the sensory root of the right fifth nerve within medulla oblongata, lesser degree of the same condition in the adjoining and closely-connected vagus root (hence supra-orbital pain, local anæsthesia and vomiting); extension of the morbid process to the motor root (hence vaso motor paralysis and secretory and trophic changes in the conea, superciliary periosteum, etc). Second period: recovery, to a large extent, of the nutrition of the posterior root of the trigeminus, complete recovery of the root of the vagus (hence alteration of the type of recurrence of the pains, which now occur at increasingly long intervals, and needed special provocation, e. g., excessive fatigue, to bring them on; hence, also, disappearance of the stomach symptoms); continuance of the affection of the motor portion of the nerve (hence, continuance of the tendency to trophic, secretory, and vaso-motor changes; develpoment of the true points douloureux during and after the paroxysms, instead of the diffused tenderness following the old attacks of migraine. Third stage: neuralgic attacks become rare and comparatively unimportant; tendency to trophic changes greatly lessened; local anæsthesia persists. Presumption, that the nutrition of the nerve-centre has nearly recovered itself, but that that centre is still the *locus minimœ resistentiœ* of the central nervous system, liable to suffer from any cause of general nervous depression.

Now, in interpreting the above phenomena, as I do, upon the theory of one essentially uniform nutritive change affecting the fifth nerve within the medulla oblongata, I shall be met with the following objections: First, there is the common and superficial difficulty that pain and paralysis of sensation must be opposite states, and that it is impossible to refer them both to one and the same pathological process. I have already in many places given instances how constantly pain and sensory paralysis interchange in a manner which is totally incomprehensible except upon the supposition that their physiological basis is essentially the same; but the most satisfactory evidence, perhaps, that could possibly be produced on this point is to be found in the perusal of a group of cases observed by Hippel,* and entitled by him "Anæsthesia of the Trigeminus," the loss of sensation being the most remarkable feature. The cases are so deeply interesting that I would gladly transfer them bodily to these pages, but must abstain from want of space. Suffice it to say here, that, in the first place, the anæsthesia was accompanied, in every one of these cases, by a most distinct and typical neuralgia; and, secondly, that trophic changes occurred which most interestingly (though not with absolute completeness) reproduced the phenomena observed after complete section of the trigeminus at the Gasserian ganglion.

The second objection sure to be raised to the theory of a simple spreading of a nutritive central change, as the cause of all the phenomena in such a case as the above, is this: It will be asked how the process extended itself to the motor root, which, in the case of the fifth nerve, is removed by a somewhat formidable anatomical distance from the sensory root. I am, of course, well aware of the latter fact, and it is an additional reason for selecting neuralgia of the fifth, as an extra difficult test of the value of my theory. A few words must be premised, reminding the reader of the physiological anatomy of the nerve.

The trigeminus is in all its characters a spinal nerve; but it has sundry peculiarities both of structure and of connections with other nerves. Its posterior or sensory root is enormous, and, as Schroder van der Kolk showed, takes a direction from behind downward and forward, which is intended to facilitate its numerous and important connections with the nuclei of other nerves: of these the most notable are its connections with the vagus, facial, glosso-pharyngeal, and hypo-glossal nuclei. The motor root, much smaller than the sensory, was shown by Lockhart Clarke to be traceable as low as the inferior

---

* "Ernährungsstörungen der Augen bei Anæsthesie des Trigeminus." Mitgetheilt von Dr. v. Hippel in Konigsberg in Preussen. Archiv f. Ophthalm. Band. xiii

border of the olivary body, as a column of cells which occupies a situation corresponding to that of the anterior course of the spinal gray matter.

As this column passes onward in the medulla oblongata, on a level with the glosso-pharyngeal nerve, it forms a group of cells of large size. Besides numerous other connections which it forms, Clarke describes the motor root as sending processes forward, like tapering brushes or tails of fibres, in connection with more scattered cells lying in their course, which may be frequently seen to communicate with the transverse bundles which traverse the "gray tubercle" and the sensory roots of the fifth contained therein. In this way the sensory root, though seemingly much separated from, is really in very direct connection with, the motor root.

Now, proofs, which must be considered almost positive, have recently been adduced to show that the nerve-fibres concerned in those peculiar alterations in the tissues supplied by the ophthlamic division of the fifth, which occur in section of the trigeminus, come entirely from the motor root of the fifth, and form a very small band in the inner or medial margin of the ophthalmic trunk. The observation of Meissner* goes to show that it is possible (by good luck) to divide the trunk in such a partial manner as to cut only the inner fibres, and thereby produce the trophic eye-changes without any anæsthesia, or only the sensory fibres, and thereby induce anæsthesia without any trophic changes; and it must be owned that this really affords the only reasonable explanation of the discrepancy between the experimental results obtained by Magendie and Bernard; and also the facts of such cases as those related by Mr. Hutchinson, † who in two instances found that a completely anæsthetic eye recovered perfectly well from the wound made in a surgical operation. The nature of the nervous influence (whether ordinary vaso-motor only, or a special trophic function) has been greatly disputed. Dr. Wegner,‡ from observing the remarkable group of glaucomatous cases under Horner (of which one has been related), made experiments, from which he concluded that the augmentation of intraocular pressure in glaucoma was a phenomenon dependent upon the sympathetic, which was irritated by reflec-

---

* Zeitsch. f. rat. Med., 1867. There is corroborative evidence, from independent sources, of the truth of Meissner's views. His own observation only proved half the case; but he quotes an observation of Buttman's in which the exact converse of his own experience happened, the external fibres being affected without the inner band, and anæsthesia without trophic changes being the result. Moreover, Schiff (Gaz. hebdom., 1867) obtained experimental results (in operating on cats and rabbits) which coincide with Meissner's.

† London Hospital Reports, vol. iii., p. 305.
‡ Wegner, loc. cit.

tion from the trigeminus. But the researches of Hippel and Grunhagen, especially their latest,* give a different explanation, excluding the sympathetic; they found that irritation of the medulla oblongata, in the neighborhood of the trigeminus root, produced a lasting and very pronounced augmentation of intra-ocular blood-pressure, an effect which, they remark, could not depend on irritation of the vaso-motor centre, since that must produce contraction of the vessels and lowering of the blood-pressure. They conclude that "the trigeminus contains specific fibres which possess the property of actively dilating the blood-vessels of the eye;" and in reference to the secretion of the fluid humors of the eye, they conclude also that "the trigeminus also plays the part of an (active) nerve of secretion."

Of these conflicting opinions I can have no difficulty in at any rate rejecting that of Wegner; for the clinical phenomena of the complications attending trigeminal neuralgia, such as they are described in my last chapter (and could have been described at much greater length), seem to me utterly to exclude vaso-motor spasm except as a temporary phenomenon at the commencement of the attacks of acute pain. Vasomotor palsy undoubtedly is very often present, in fact every attack of neuralgia of a certain severity is thus complicated; and there is no reason to doubt that this paralysis could be caused by lesions within the medulla. Are we, then, to admit functions of active dilatation of vessels, and active impulse to secretion in certain fibres of the fifth ? It is necessary at any rate to clear the ground in one respect: it must not be supposed that I for a moment entertain the idea that there can be direct active dilatation, *i. e.*, that there can be any system of muscular fibres (and nerve-fibres stimulating them) whose office is to open the calibre of the vessels; the idea is wildly improbable—in fact almost inconceivable by any one who reflects on the necessary machinery—and there is not a single observed anatomical fact to give it support. If, then, I speak of the possibility of "active" dilatation, it must be understood that I refer to a theory of "inhibition," which supposes certain fibres to be gifted with the power of paralyzing or inhibiting the vaso-motor nerves. It is my duty to speak with all reasonable reserve on that most difficult *quæstio vexata*, the existence of special inhibiting systems of nerves, and the extent to which a double series of opposed nervous actions is generalized in the body; but it is impossible to avoid the subject altogether, and I offer the the following remarks, with deference, to our professional physiologists. The strongest instances of the apparent inhibiting action are probably afforded by the *nervi erigentes*, as shown by Loven, the cardiac depressor, by Lud-

---

* Archiv f. Ophthalm., xv., 1,

wig and Cyon, and the splanchnics (upon the intestine), by Pfluger. But there is not a single one of these examples that has not been challenged by experimenters of repute. Thus the theory of the distinctive restraint-action of the splanchnics upon the intestine, and of the vagus upon the heart, has been especially controverted by Piotrowski, who, indeed, rejects the whole theory of special inhibitory nerves.* And, from another point of view, Mr. Lister long ago attacked the views of Pfluger, maintaining that it was possible to produce exactly opposite effects through the medium of the very same nerves, according as the experimental irritation applied to them was weak or strong. To Dr. Handfield Jones † this seems a still unanswerable objection to the inhibitory theory. And in the remarkably able and judicial summary of the "Physiology and Pathology of the Sympathetic or Ganglionic System," ‡ by Dr. Robert T. Edes, a less decided but still tolerably strong acquiescence is given to Mr. Lister's criticisms of this theory. Personally, I must express very strongly the distrust (which is probably felt by many others) of doctrines which assert an exact opposition between the functions of any two nerves, on the basis of an observation that the same apparent effects may be produced by section of the one and galvanization of the other; both processes seem far too pathological, and too remote from the conditions of ordinary vitality, to admit of any such absolute deductions from their results.

In the present state of our information I am inclined to explain all the congestive complications of trigeminal neuralgia on the basis of vaso-motor paralysis. And I further believe that the cause of that paralysis is a direct extention of the original morbid process from the sensory root to the motor, affecting the origin of fibres in the latter, which are destined to govern the calibre as ocular and facial vessels. These fibres I suppose it is that Meissner succeeded in dividing when he partially cut the trigeminus, and got nutritive and vascular changes without anæsthesia.

There must be more than this, however, to account for the whole of the trophic phenomena; for there is a great body of evidence to show that mere vaso-motor paralysis does not produce any phenomena of such an actively morbid kind as those we are endeavoring to explain. The phenomena on the side of secretion might indeed be possibly explained by vaso-motor paralysis. [It must be remembered that I am speaking of such

---

* "Deutsches Archiv f. klin. Med.," ii., 2, 1866. I am not aware whether Piotrowski has at all altered his opinions since the (subsequent) observations of Ludwig and Cyon upon the "depressor" nerve.
† "Functional Nervous Disorders." Churchill, 2d edit., 1870.
‡ "Prize Essay of the New York Academy of Medicine." New York: Wood & Co., 1869.

augmented secretion as is seen in neuralgia. I agree with Prof. Rutherford (Lectures on Experimental Physiology, Lancet, April 29, 1871) that it is difficult thus to explain the effects of galvanization of the chorda tympani on the sub-maxillary gland.] Consisting as they do (a), in the great majority of cases, of a mere outpour of what seems little more than the aqueous part of the secretion, and (b) in a few cases of arrested secretion, a phenomenon otherwise by no means unfamiliar as the result of sudden, passive engorgment of glands. But the mere cessation of vaso-motion will not account for such facts as the rapid and simultaneous development of erysipelatous inflammation, of corneal clouding and ulceration, of iritis and glaucoma, of nutrition-changes in hair and mucous membrane. I must, for the present, be content to believe it probable that there is a special set of efferent fibres in the trigeminus, emanating from the motor-root, whose office it is in some unknown way to preside over the equilibrium of molecular forces in the tissues to which the nerve is distributed; trophic nerves, in fact, though not active dilators of blood-vessels.

It seems to me that, without enlarging further on this almost endless topic, I should be justified in assuming that I had shown the very high probability that the common starting-point both of the neuralgia and of its vaso-moter secretory, and trophic complications, was in the sensory root of the trigeminus. But the argument is greatly strengthened when we consider the fact that loss of peripheral common, and also tactile sensation, to a greater or less degree, is constantly observed to occur simultaneously with the pain and with the other complications. When we observe a patient suffering from racking supra-orbital and ocular neuralgia, and discover that at the very same period the skin round the eye is markedly insensitive to impressions, except in the *points douleureux*, what can we rationally suppose, except that both pain and insensibility are the result of one and the same influence, which radiates from the sensory centre ?

Nor are we likely to reach a different conclusion, if we test the matter by the consideration of a rarer, but still sufficiently common kind of case, such as I have described in Chapter I., in which a very strong peripheral influence (traumatic) produces neuralgia, accompanied by vaso-motor and secretory phenomena, and by anæsthesia, but not in the district of the painful nerve, but in the territory of a quite different nerve. How can we doubt, in the case, *e. g.*, of a trigeminal neuralgia thus complicated, the exciting cause of which was a wound of the ulnar nerve, that the morbid influence, traveling inward from the lesion, would have passed without any special consequences (as happens in thousands of such nerve-wounds), had it not, in its passage along the medulla, encountered a *locus minoris resistentiæ* in the roots of the trigeminus ? It seems impossible to account for the phenemena on any other theory. [Eulenburg says, in reference to my reported cases of the kind: "*Solche Fälle begunstigen in hohem Grade die Annahme pradisponirender Momente, die in der ursprunglich schwacheren Organisation einzelner Abschnitte des centralen Nerven-apparates beruhen.*" *Op. cit.*, p. 56.]

It is necessary, in the next place, to consider a very important question, how far irritation can pass over from one nerve to another, without reflection through a spinal centre, solely in virtue of a connection through the medium of a nervous plexus. The case which apparently presents such phenomena in the most unmistakable way is that of *angina pectoris*.

The site to which the essential heart-pain is referred in this disease is probably the cardiac, or this and the aortic plexus; in a comparatively small number of cases the pain does not extend farther. But much more frequently it spreads in various directions, and we have to account for its presence (*a*) in intercostal nerves, (*b*) cervical nerves, (*c*) nerves springing from the brachial plexus.

Before we inquire into the mechanism by which this extension of the pain takes place, we ought in strictness to ask ourselves whether the essential heart-pain is felt only in the spinal sensory branches, or whether the sympathetic fibres are themselves capable of feeling pain. The latter supposition, notwithstanding all that has been argued in its favor from the supposed analogies of the pain of colic, gall-stone, etc., seems to me very doubtful. It would appear more probable that both the latter pains, and also those of angina, are really connected with branches either of the vagus or of other spinal nerves. And there is no need to invoke the sympathetic as a sensory nerve, to account either for the essential heart-pain of angina, or for its extension into arm, chest-wall, and neck. For the plexus cardiacus receives spinal branches, both from the vagus and also (through the medium of the sympathetic ganglia of the neck) from the whole length of the cervical and the uppermost part of the dorsal cord-centres. And, in this way, it would seem quite possible intelligibly to account for the pain radiating into intercostal, cervical, and brachial nerves, merely by extension of a morbid process essentially seated in the cord. Usually, however, one sees it explained not in this way, but by the inter-communications that exist outside the spine, between the branches from the cervical ganglia and the lower cervical and upper dorsal nerves; and the pain in the arm is especially explained by the connection (outside the spinal canal) of the inferior cervical ganglion, on the one hand with the lower cervical nerves, which go to the brachial plexus, and, on the other hand, with the heart itself. There remains to be explained, however, the singular tendency of the arm-pain to be one-sided (this happens in at least four cases out of five); and this explanation seems to me insuperably difficult, on the theory that the transference of morbid action to the brachial nerves takes place through external anastomoses. It appears greatly more probable that angina is essentially a mainly unilateral morbid condition of the lower cervical and upper dorsal portion of the cord; liable of course to be seriously aggravated by such peripheral sources of irritation as would be furnished by diseases of the heart, and especially by diseases of the coronary arteries; the latter affection probably involving constant mechanical irritation of the cardiac and the aortic plexuses. It is noteworthy that the arm-pain is sometimes (I do not know how often) accompanied by vaso-motor paralysis in the limb; this phenomenon could also certainly be more easily accounted for on the supposition of radiation from a spinal vaso-moter centre (to which the morbid process had extended from a posterior nerve-root) than on that of communication between painful sensory nerves and vaso-motor nerves; through either of the plexuses independently of the spinal centres.

In truth, I suspect that, whatever part the plexuses, with their reenforcing ganglionic cells, may play during physiological life, they are not often the channels of mutual pathological reaction of one kind of nerve with another. It would be possible to argue this even more strongly in the case of trigeminal neuralgias; but I must not unnecessarily expand this already too lengthy discussion.

From the varied considerations which have now been adduced, the reader, unless I altogether miscalculate the value of the facts, will probably have arrived at the following conclusions: (1) That the assumption of a positive material centric change as the essential morbid event in neuralgia is almost forced upon us; (2) that, whereas the morbid process, if centric, is *a priori* infinitely more likely to be seated in the posterior root of the painful nerve, or the gray matter immediately connected with it, than anywhere else; so, again, the assumption of this locality will explain, as no other theory could explain, the singular variety of complications (all of them nearly always unilateral, and on the same side as the pain) which are apt to group themselves around a neuralgia; and some of which are very seldom absent in neuralgia of any considerable severty. To this we may certainly add that it is extremely probable that the vast majority of neurlgic patients inherit the tendency to this localized centric change; in support of this we may finally mention two considerations derived from the sex and the ages most favorable to neuralgia. Eulenburg saw a hundred and six cases of neuralgia of all kinds, of which seventy-six were in women and only thirty in men; my own experience is very similar; namely, sixty-eight women and thirty-two men out of a hundred hospital and private patients. The strong connection between the hysteric and the neuralgic temperament in women, and the great preponderance of women among neuralgics, strengthen in no small degree the probability of inherent tendencies to unstable equilibrium as a very common predisposing factor in neuralgia. And, on the subject of age, I need only recall what I have said so strongly about the coincidence of neuralgia with particular epochs in life, as affording evidence of the most powerful kind that neuralgics are, save in exceptional instances, persons with congenitally weak spots in the nervous centres, which break down into degeneration, temporary or permanent, under the strains imposed by one or other of the physiological crises of the organism, or the special physical or psychical circumstances which surround the patient's life.

Having thus decidedly expressed by belief in the essential material participation of the nerve-centre in neuralgia, it remains for me to discuss two points: first, as to the character of the material change in the nerve-root, and next, as to the extent to which mere peripheral influence, without special inherited tendencies, may suffice to set this process going.

The morbid change in the nerve-centre is probably, in the vast majority of cases, an interstitial atrophy, tending either to recovery, or to the gradual establishment of gray degeneration, or yellow atrophy, of considerable portions of the whole of the posterior root, and the commencement of the sensory trunk as far as the ganglion.

It is probable, however, that in a certain number of cases, the atrophic stage may be preceded by a process of genuine inflammation, and that this inflammation is centripetally produced in consequence of inflammations of peripheral portions of the nerve. The considerations which make this probable are chiefly derived from the analysis of cases in which a more or less chronic, but severe, visceral disorder has been followed by so-called reflex paralysis, but in which neuralgic phenomena have been conspicuous. In reference to this subject I recommend to the reader's attention the very interesting paper on "Reflex Paralyses" by Prof. Leyden, of Konigsberg.* He is immediately commenting upon a case in which dysenteric affection of the bowel were followed by the symptoms of myelitis, attended with febrile exacerbations, and also with severe pains in the region of the sacrum, in the course of the dorsal intercostal nerves of the right side, and in the knees, and semi-paralytic weakness of the lower extremities, and with pains between the shoulder-blades and the left arm. Leyden discusses the doctrine of reflex paralyses in general, starting from

---

* Volkmann's Sammlung klinischer Vortrage, No. 2. "Ueber Reflex Lahmungen," von E. Leyden. Leipzig, 1870.

the cases of urinary paraplegia brought forward by Stanley, in 1835, and tracing the growth of opinion through the phases represented by Graves, Henoch, and Romberg, by Valentine and Hasse, then by Pfuger, and other professors of the inhibitory doctrine; by Brown-Sequard (in his well-known, and now very generally discredited, theory of spasm of the vessels in the nervous centres), by Jaccoud in the "Erschopfung" (exhaustion) theory, down to the more careful and reliable researches of Levisson on the temporary reflected paralyses induced by experimental squeezing of the kidney or uterus of animals; and then gives the history of the more recent doctrine of a positive material change in the cord centripetally introduced. Gull[*] (1856) may be said to have inaugurated the new doctrine of a morbid process transmitted along the pelvic nerves to the cord, and causing material changes there. Remak,[†] on the other hand, suggested a material change operating in the opposite direction; *a neuritis descendens*, starting in the very nerves (within the pelvis) which showed the paralysis in the extremities. The symptoms are supposed by him to be distinctive, inasmuch as there is both violent pain in the nerves of the soles of the feet, and also tenderness of the same. On the other hand, Remak said that myelitis, with neuritis, might be the origin of paraplegia and simultaneous palsy of bladder and rectum. The theory of neuritis descendens was supported by Kussmaul,[‡] in the record of a case where disease of the bladder was complicated with pelvic inflammation, atheromatous degeneration of the arteries, and consequent fatty degeneration of the sciatic nerves, causing direct paraplegia. We return to the centripetal theory of urinary paralysis with Leyden's own cases, published in 1865; of three patients with urinary paraplegia, two died, and the existence of a secondary (centripetal) myelitis seems to have been established, and by all analogy it must have existed in the third case, which recovered. The only puzzle and doubt that ensued was caused by the fact that there was an absence of neuritis in the different nerves themselves; though it seemed plain that the starting point of the myelitis was at the entrance of these nerves into the cord. This mystery seemed to be cleared up by the important experiments of Tiesler, ("Ueber Neuritis" Konigsberg, 1860) a pupil of Leyden's. This observer excited local traumatic inflammation in the sciatic nerve of rabbits and dogs; the rabbit became paraplegic and died three days afterward. At the site of the artificial irritation there was a localized formation of pus, and there was a second similar formation within the vertebral canal at the point where the posterior

---

[*] "Cases of Urinary Paraplegia," Med.-Chir. Trans., 1856.
[†] Wurzburg. Med. Zeitsch., iv., 56–64.
[‡] Med. Cent. Ztg. 21, 1860.

roots of the sciatic enter the cord; but there was no neuritis of the intervening portion of the nerve.

Upon this and similar evidence is based the modern doctrine of a neuritis migrans, with centripetal tendencies, upon which it is supposed that a very large proportion, at least, of the urinary, dysenteric, and uterine paraplegias, miscalled "reflex," depend; and it ts clear that the application of the word "reflex" in such a case is a grave abuse, tending to produce such confusion of thought and error in practice. In relation to the subject of our own inquiry—neuralgia—it is obviously of the highest consequence to investigate the qnestion whether peripheral irritations, analogous to those which produce urinary paraplegia, are at all frequently the cause of the changes in the posterior roots which produce true neuralgia; for of course an inflammation may be the beginning of an atrophy which may presently exhibit no distinction whatever from one of which the origin was altogether non-inflammatory. I think that there is strong reason for thinking that this is not at all frequently the case. In the first place, all the evidence that exists respecting these centripetal inflammations of the cord is opposed to the idea that, save in the rarest instances, the inflammatory process limits itself to one small segment of the cord. Secondly, the description of the pains that have usually accompanied such inflammations of the cord is considerably different from the strictly localized, frankly intermittent character of a true neuralgia; in fact, all we know of the history of myelitis (except when complicated with a large amount of meningitis) forbids us to suppose that severe pain would be an immediate symptom. But, thirdly, a far more important objection to the theory of an origin in localized centripetal myelitis, the result of a neuritis migrans, is the rarity of motor paralysis as an early symptom, instead of which we ought to find a very distinct history of decided paralysis (much more decided than those secondary paralyses which actually do occur in some neuralgias) of the muscles supplied by the anterior roots of the painful nerve, in every case in which such a peripheral origin could be assumed. Again, the totally feverless commencement of neuralgias, a character which is maintained throughout the progress of the milder cases, is entirely opposed to the idea of a direct connection between myelitis and neuralgia. The superficial appearance of pyrexia is sometimes given by a local vaso-motor paralysis. which makes the neuralgic part, after a long bout of pain, hot and red; but of general pyrexia there is nothing.

Taking every thing into consideration, one is inclined to say that there is a probability that in a very limited number of cases peripheral irritation does cause actual limited myelitis, which escapes recognition at the time, but which issues in an atrophy, the subjective expression of which is actual neuralgic pain. We may well ask ourselves, also, whether there is not some likelihood that a peripheral irritation, which stops short of producing an actual neuritis migrans capable of centripetally exciting a myelitis, may not, by a lower degree of centripetal irritation, give a bias toward certain forms of non-inflammatory atrophy in cells of posterior nerve-roots which are congenitally of weak organization. I am inclined to believe strongly that this does occur. For example, I should explain thus the majority of the peripheral cases of ciliary neuralgia, migraine, etc., that we meet with in poor young needle-women, especially the hypermetropic, who, at an age when they can ill afford the strain, work so constantly and strenuously at an occupation which fearfully taxes the eye.

I would also go farther, and express the opinion that peripheral influences of an extremely powerful and continuous kind, where they occur with one of those critical periods of life at which the central nervous system is relatively weak and unstable, can occasionally set going a non-inflammatory centric atrophy which may localize itself in those nerves upon whose centres the morbific peripheral influence is perpetually pouring in. Even such influences as the psychical and emotional, be it remembered, must be considered peripheral—that is, they are external to the seat and centre of the neuralgia. And there are probably few practitioners of large experience who have not seen a patient or two in whom the concurrence of some unfortunate psychical with some other noxious peripheral influence, the whole taking place at some critical period of life (especially in the years between puberty and marriage), seems to have totally deranged the general balance of nervous forces, and induced morbid susceptibilities and morbid tendencies to some particular neurosis. It is a comparatively frequent thing, for example, to see an unsocial solitary life (leading to the habit of masturbation), joined with the bad influence of an unhealthy ambition, prompting to premature and false work in literature and art. The bad peripheral influence of constant fatigue of the eyes in study may so completely modify a young man's constitution as to make a wreck of him in a very few years, changing him from the state of habitual and conscious health to that of chronic neurosis of one sort or another. And, though it is doubtless on persons with congenial tendencies to nervous diseases that such a combination of bad influences produces its most serious effects, yet there unquestionably are a few persons in whom they appear to entirely generate the neurotic constitution. I have already touched upon the part that misdirected psychical influences, especially religious and other forms of emotional excitement, may play in this unfortunate perversion of the natural and healthy nervous functions, more especially in youth;

and need only add, here, that perhaps the most fatal combination of all the bad influences is the melanchoy union of highly-strained religious sentiment with peripheral sexual irritation, which is, unfortunately, a too common phenomenon under certain systems of education. The most frequent neurotic consequences of the class of influences which have now been referred to are probably neuralgia—in the form either of migraine, of nervous angina, or of sciatica—or else asthma.

But, if the combination of several such centripetal influences may generate the neurosis unaided, even a single one of them operating powerfully for a long period may produce most serious consequences in those who are hereditarily predisposed. The influence of prolonged fatigue of the eyesight, independently of any special intellectual or emotional strain, was strongly illustrated in my own case about three years ago. I was then engaged upon a piece of scientific writing which demanded no great intellectual effort, but was being done against time, and by working, night after night, many hours by gas-light. My neuralgic (trigeminal) attacks came on with great severity, accompanied by vertiginous sensations of so alarming a kind as to make me fear the invasion of some serious brain-mischief. I broke off all work, and went to the sea-side, but was greatly disappointed to find, for the first few days, that the symptoms were not in the least mitigated. The mystery was soon explained. The weather had been such as to confine me a good deal to the house, and, thinking it would do no harm, I amused myself with reading newspapers and novels. At last I suspected that the use of my eyes in reading was altogether mischievous; I desisted from reading anything, and in forty-eight hours every symptom had vanished.

Among peripheral influences of a more mechanical kind there is one cause of neuralgia, the force of which has been variously estimated, but which some authors rate as very important, viz.: the influence of the pressure, and especially of the varying pressure, of blood-vessels, or other hollow viscera, upon the trunks of the nerves. We must set aside one such action which is undoubtedly very powerful, as essentially differing from the others; I mean the pressure of dilated blood-vessels, especially aneurisms, when this happens to be exerted upon the ganglion of the sensory trunk. Here there can be no doubt of the mischief; for the pressure, if at all severe, gradually destroys the life of the ganglion, upon which, as was proved by Waller, the nutrition of the posterior nerve-root hangs with very intimate dependence, and the pulsations of the vessel seem greatly to aggravate both the irritation and the centripetal tendency to atrophy. In short, it is plain that such lesion of a ganglion may be the whole and sufficient cause of a neuralgia of the most desperate and incurable kind. It is another matter when we are asked to believe that the mere varying pressure of intestines, in different states of fullness, or plexuses of pelvic veins liable so temporary congestions, can so affect the sciatic nerves as to set up neuralgia. Considering the extreme frequency of cases in which such momenta must be partially coming into operation, especially in women—a frequency altogether out of proportion to that of sciatica—I cannot admit the probability that this influence is more than an occasional and very secondary factor, and that only in cases where the disposition to neuralgia is uncommonly strong.

A sufficiently complete explanation of my theory as to the pathology and etiology of neuralgia has now been given, although the subject might be elaborated at far greater length; and I hope it will be apparent to the reader that the view now advocated is at once important, and also vouched for by strong evidence. I claim for it that the whole argument shall be taken together, for it is a case of cumulative proof; every link must be weighed and tested, before the remarkable strength of the chain can be felt. And it may fairly be said that, if the proof of a definite kind of material change in a definite organ, as the essential factor in neuralgia, has been established upon reasonable grounds, an important step has been taken toward removing a serious opprobrium and difficulty in practical medicine. Although the true neuralgias are not among the most frequent of human diseases, they form a class of enormous practical importance, for they are sufficiently common to be sure to occur in considerable numbers in the practice of every medical man, and, both from the suffering which they inflict, and the rebelliousness which they often show to treatment, they are among the gravest sources of anxiety which the practitioner is likely to encounter. There are probably few disorders which so often occasion mortification and loss of professional credit to the physician. The helplessness which men, who do not enjoy special opportunities of seeing those diseases with frequency, so often show in dealing with them, is largely caused by the extreme timidity and vagueness with with which the standard treatises on medicine deal with the question of their pathology; and a very unfair advantage has thus been given to the specialists, who, by the mere force of opportunity, and continual blind "pegging away" in an entirely empiric manner, have acquired a certain rude skill in the treatment of these maladies which enables them to outshine practitioners who often have far more in them of the veritable *homme instruit* as regards general scientific education and habits of mind. It will be evident, as a mere abstract proposition, that the enunciation of a reasonable pathology of the disease, and the sweeping away of a mass of unmeaning phrases about "mysterious functional affections" and the like, must be a distinct gain to practitioners of plain commonsense and good general knowledge, to whom neuralgia is merely one of a vast number of different diseases among which their attention and study are divided. And I hope that, in the further remarks on Diagnosis, Prognosis, and Treatment, yet to be made, the value of clear pathological ideas of disease will be brought more practically and clearly into view. [The reader will find, at the end of Part I. of this volume, a note which contains a brief discussion on the "Erschopfung" theory of Jaccoud, and the doctrines of Dr. Handfield Jones respecting inhibition, with which I thought it best not to encumber the text of the present chapter.]

CHAPTER IV.

DIAGNOSIS AND PROGNOSIS OF NEURALGIA.

Diagnosis.—This subject is much simplified and shortened, in regard to our present purpose, by the plan of the present work, which, by separately describing (in Part II.) the other disorders which resemble neuralgia, and are liable to be confounded with it, avoids the necessity for stating here the negative diagnosis of neuralgia itself. We are only concerned here to give a clear picture of the positive signs which it is necessary to verify before we can suppose disease to be neuralgia. The special modes of searching for these are interesting, and in some respects peculiar:

(1) The first and most essential characteristic of a true neuralgia is, that the pain is invariably either frankly intermittent, or at least fluctuates greatly in severity, without any sufficient and recognizable cause for these changes.

(2) The severity of the pain is altogether out of proportion to the general constitutional disturbance.

(3) True neuralgic pain is limited with more or less distinctness to a branch or branches of particular nerves; in the immense majority of cases it is unilateral, but when bilateral it is nearly always symmetrical as to the main nerve affected, though a larger number of peripheral branches may be more painful on one side than on the other.

(4) The pains are invariably aggravated by fatigue or other depressing physical or psychical agencies.

The above are characteristics which every genuine neuralgia possesses, even in its earliest stages; if they be not present, we must at once refer the diagnosis to one or other of the affections described in Part II. of this work.

Supposing the above symptoms to be present, we expect to find—

(5) In by far the largest number of instances that the patient has either previously been neuralgic, or liable to other neuroses, or that he comes of a family in which the neurotic disposition is well marked. Failing this, we are strongly to doubt the neuralgic character of the malady, unless we detect that there has been—

(6) A poisoning of the blood by malaria (but this very rarely causes neuralgia, save in the congenitally predisposed); or—

(7) A powerfully operating or very long-continued peripheral irritation centripetally directed upon the sensory nucleus of the painful nerve; which irritation may be (*a*) "functional." as where the eye has been persistently and severely overstrained and trigeminal pain results, or a sudden severe shock has been received; or, (*b*) coarsely material, as where inflammation, ulceration, etc., of surrounding tissues involve the periphery of the painful nerves in a perpetually morbid action, or chronic but profoundly depressing psychical influences; or—

(8) A constitutional syphilis. In this case there will either be marked syphilitic local affection of the trunk of a nerve, or if, as is more common, the syphilitic change is in the nerve-centre, there will most likely be other syphilitic centric mischiefs, leading to scattered motor or vaso-motor paralyses, characteristic modifications of speciel sense-functions, etc.

If the neuralgia be of some standing and a certain degree of severity, there will inevitably be found—

(9) Some of the fixed tender points of Valleix, in such situations as have been described in Chapter I.; and—

(10). Secondary affections (*a*) of secreting glands, or (*b*) vaso-motor nerves; or (*c*) of nutrition of tissues; or secondary localized paralyses of muscles, or localized anæsthesia of a somewhat decided though not complete kind, as described in Chapter II.; any one or any number of these various complications may be present.

I must insist that the above picture includes only the essentials for a diagnosis of neuralgia; if the painful affection will not answer to the conditions therein included, we have no right to call it a neuralgia—it belongs, for every practical purpose, to some other category of disease. Let me add one more essential characteristic, which is, that the pain begins and assumes its characteristic type before any other of the phenomena appear, with the single and partial exception of anæsthesia.

There are some special modes of diagnosis of the varieties

of neuralgia, developed of late years, that require notice here; they are chiefly the result of the researches of Moriz Benedikt.

As regards the quality of the pain, Benedikt says that the curve of intensity has an intimate relation to the *locus in quo* of the neuralgia (*i. e.*, whether in the periphery, trunk, or roots). An inflammatory irritation set up at the periphery of a nerve (by a joint-inflammation, for instance) produces a continuous pain; the same kind of irritation, attacking a nerve-trunk (*e. g.*, in the bony canals), produces a paroxysmal pain; an inflammation spreading from the vertebræ to the nerve-roots or the cord-centres produces momentary lancinating pains. The latter characteristic he supposes to be especially characteristic of the centrally-produced neuralgias; and I may observe, as so far confirmatory of this idea, that this is especially the character of the pains in locomotor ataxy. There are sundry special cases to be considered, however: thus, Benedikt himself remarks that the pain set up by the pressure of a pulsating aneurism is, from the nature of things, lancinating from moment to moment. Eulenburg,* moreover, says that Benedikt's tests of the locality of the primary mischief only hold good under the following circumstancs: (1) When the irritability and the exhaustibility of the nerves are in a normal condition during the neuralgia; (2) when the irritation that calls forth the paroxysm is either identical with the original cause of the disease, or at least operates upon the same spot. The two conditions, however, do not concur. The irritability and exhausibility may be sometimes excessive in neuralgias, sometimes normal, and perhaps, in certain cases, beneath the normal standard; by which means the form of the curve of intensity must be considerably modified. Moreover, the irritation that provokes an attack may from the periphery attack the primary seat of the disease, even when this is central, on account (says Eulenburg) of exaggerated conductivity of the nerves (his second cause† of "hyperæsthesia"), as is, in fact, very frequently the case. He also thinks the distinction between paroxysmal and lancinating pains too indefinite to serve as a sufficiently reliable basis of diagnosis, especially considering the endless *nuances* of the form which the pain is apt to take. I agree with Eulenburg upon this point; and am convinced, from my own observations, that such a distinction as that between lancinating and paroxysmal pains is illusory, [I have taken some pains to investigate the character of the pains, not only in neuralgia, but in locomotor ataxy. It is true that the lancinating character predominates, on the whole, in the latter disease; but there are great differences in different individuals, and even in the same patient at various

---

* *Op. cit.*, pp. 65, 66.
† Idem, p. 8.

times, which plainly depend on subjective influences. Compare, for instance, Dr. Headlam Greenhow's report on an ataxic patient, with a report on the same man by Dr. Buzzard and myself. ("Trans. Clin. Soc.," vol. i., 1868, pp. 152–162.)] the two kinds being frequently found alternate in the same case. The only useful distinction, in my opinion, is Benedikt's first one: he is probably right in saying that, where such an affection as an inflamed joint forms the source of peripheral irritation that immediately provokes a neuralgia, the pain is apt to be unusually continuous.

The extent to which the pain of neuralgia spreads into different termini of the same nerve has been made the basis of distinctions as to the seat of the original mischief. For example, it has been said that pain in the mental branch of the third division of the trigeminus, which does not invade the auriculo-temporal branch, can hardly depend on an irritation operating on the trunk of the inferior dental; it must be distinctly peripheral, or else it must act upon limited portions of the central origin of the fifth nerve. But the fact seems rather to be that, whether the neuralgia was excited by lesions at the periphery, in the nerve-trunk, or in the centre, it is equally possible that either a small or a large part of the peripheral expanse of the nerve may become the seat of the pain: this almost necessarily follows from the entire independence of individual fibres in nerves.

As regards the evidence afforded by the motor, vaso-motor, and trophic complications, there is this very positive diagnostic value in them—that they enable us to say, with greater assurance than we could otherwise do, that the disease is a real neuralgia. But, the only evidence that they afford as to the situation of the mischief is, that they uniformly point to the central end of a particular nerve; and accordingly I have already shown, in the chapter on Pathology, that the attentive study of these very complications furnishes us with some of the most powerful arguments upon which rests my theory that in neuralgia there is always centric mischief. What share in the production of the malady, in any given case, has been taken by the centric disease, and what if any by a peripheral irritation, the existence of these complications in no way helps us to determine; far less does it enable us to localize a peripheral lesion which may have acted as a concomitant cause; on the contrary, I believe that there is no more fertile source of erroneous judgment on this very point, than some of these complications, especially the vaso-motor and trophic. I suspect that it has happened, in hundreds of instances, that a localized congestion or inflammation, which is a mere secondary phenomenon, produced in the centrifugal manner already so fully explained, has been taken for the veritable *fons et origo* of the malady: hence the neuralgia has been confidently reckoned as one peripherally produced, and, what is even worse, the whole energy of treatment has been directed to a mere outlying symptom, under the idea that the primary source of mischief was being attacked.

The application of electricity as a test of the nature of a neuralgia has been employed by Benedikt,* who lays down certain laws as the result of his researches. He says that (*a*) in idiopathic peripheral neuralgias the nerves are not sensitive to the current; (*b*) in neuralgias dependent on neuritis or hyperæmia of the nerve-sheath there is general electric tenderness of the nerve; (*c*) in cases where the pain has been set up by morbid processes in tissues surrounding the nerve, there is electric tenderness only at the site of these changes. I may, in general terms, express concurrence in these statements; but I must add that, as diagnostic rules they apply only to the early stages of neuralgia; for the occurrence of secondary complications may and does altogether change the condition of electric sensitiveness. It need hardly be said that the above remarks on diagnosis apply for the most part only to the superficial neuralgias, which, however, include an immense majority of the cases of neuralgias. The diagnosis of visceral neuralgias is, it need hardly be said, in most cases, a far more difficult and complicated matter. In these diseases we have often little more to guide us, in the actual symptoms, than (*a*) the intermittence of the pain, and (*b*) the absence of commensurate constitutional disturbance, especially the complete freedom from sense of illness in the intervals between the pains. We shall be obliged to rely greatly on such historical facts as the presence or absence of neurotic tendencies in the patient and his family; the possibility of his having been exposed to blood-poisoning (*e. g.*, from malaria or chronic alcoholic excess, or extreme over-smoking); the circumstance that he has been habitually overworked, or greatly exposed to agitating psychical influences; perhaps that he has been subject to a combination of several of these morbific momenta. To say truth, the diagnosis of visceral neuralgias must, at the best of times, be a difficult and anxious matter, and we can hardly ever thoroughly satisfy ourselves until we have procured some decided results from treatment; fortunately, however, it happens tolerably often that we can do this, and sometimes in a very striking way.

Prognosis.—The prognosis of neuralgia varies exceedingly, according to the form and situation of the disease, and many other considerations. There are, of course, in the first place, certain neuralgias in which the prospect is perfectly hopeless as to cure; such are the cases in which the nerve is involved in a continuously growing tumor (especially within a rigid

---

* "Elektrotherapie." Wien, 1868.

cavity, like the skull), or a slow but persistent ulcerative process.

Supposing, however, that the case is none of these, the very first prognostic consideration is that of age.

Of the neuralgias of youth, the majority either disappear altogether after a first attack, or recur a certain number of times during some years, the neuralgic tendency either disappearing or becoming greatly mitigated when the process of bodily consolidation is over. In another group the neuralgic tendency is never lost, but the form of the attacks changes, and there is far less spontaneity in the manner of their production. It is exceedingly common to see delicate boys and girls between puberty and the age of eighteen or twenty, attacked with typical migraine, which recurs regularly every three or four weeks for perhaps two or three years, then ceases to occur at regular periods, then loses the tendency to stomach complication; and, by the age of twenty-five or somewhat later, has left, as its only relic, a tendency to attacks of ophthalmic neuralgia, which come on when the patient is excessively fatigued, or encounters the close air of a theatre, or undergoes an unusual strain of mental excitement or anxiety, etc.; but which never come on without some such special provocation. So, again, there is a variety of sciatica which belongs mainly to the period between puberty and the twenty-fifth to thirtieth year, and which seems really to belong, pathologically, to the age of unsettled and irregular sexual function, the tendency to it usually disappearing after the patient has settled down happily in married life. Ovarian and mammary neuralgia have very commonly a similar history.

On the other extreme we find the neuralgias of the period of bodily decay: these are of very bad prognosis. A neuralgia which first develops itself after the arteries and capillaries have begun to change decidedly in the direction of atheroma is extremely likely, even if apparently cured for a time, to recur again and again, with ever-increasing severity, and to haunt the patient for the remainder of his days. It therefore becomes exceedingly important, in a prognostic point of view, to assure ourselves as soon as possible whether this arterial degeneration has decidedly commenced; and for this purpose I am in the habit of insisting to pupils on the great importance of sphygmographic examination for all neuralgic patients who have passed the middle age. Where we get the evidence which is furnished by the formation of a distinctly square-headed radial pulse-curve, even though there be no palpable cord-like rigidity of superficial arteries, we are bound to be exceedingly cautious of giving a favorable prognosis.

In women the period of involution of the sexual apparatus forms a crisis which, in regard to neuralgias, is of great prognostic importance. On the one hand, if the general vital status be good, and the arterial system fairly unimpaired, we may look to the completion of the process of involution as a probable time of deliverance from neuralgic troubles that have hitherto beset a woman; we know that she will probably suffer a temporary aggravation of her pains, but we hope to see her lose them altogether. On the other hand, if it should happen that she enters on the period of sexual involution with her general nutrition considerably impaired and her arterial system decidedly invaded by atheroma, it is only too likely that neuralgias recurring now, or attacking her for the first time, will assume the worst and least manageable type.

Of almost or quite equal importance with the question of the physiological age of the patient is that of his personal and family history with regard to the tendency to neuralgia and to other severe neuroses. Upon this subject I have dwelt so very fully in other parts of this work, that it is merely necessary here to repeat, that the balance of chances is most heavily swayed to the bad side by all evidence tending to prove congenital neurotic tendencies in the patient and vice versa.

Of prognostic hints that are to be gathered from our knowledge of the immediate causes of the attack, there are none so valuable as those which we gather from the detection of a malarial or a syphilitic factor in the production of the malady. In the former case, we hope to cure the patient either with quinine or arsenic, with almost magical certainty and rapidity; in the latter, we expect an almost equally brilliant result from iodide of potassium.

The particular nerve in which the neuralgia is seated does not so decidedly influence the prognosis, according to my experience, as is stated by some authors; nevertheless, there are differences of this kind. For instance, sciatica, though by no means so frequently a mild and trifling complaint as Eulenburg would make it to be, is certainly, on the whole, more curable than the trigeminal neuralgias taken as a group. I, however, cannot share Eulenburg's opinion as to the rarity of a central cause for sciatica, nor his consequent explanation of its more frequent curability; the latter I explain by the fact that it is possible far more completely to remove the concomitant causes in sciatica than in trigeminal neuralgia. By simply keeping a sciatic patient in the prone posture, shielded from cold and from pressure on the nerve, we have it in our power to remove nearly all peripheral sources of irritation; but in trigeminal neuralgia there are many influences, particularly psychical ones, which cannot be shut out, and which will continue to act with disastrous effect in many cases. With all this, however, we see a sufficiently large number of incurable sciaticas, on the one hand, and of severe trigeminal neuralgia cured on the other. It is only the genuine epileptiform tic, occurring in subjects whose arterial system is an advanced stage of degeneration, that stands out clearly and unmistakably pre-eminent among neuralgias for rebelliousness to treatment of every kind.

CHAPTER V.

TREATMENT OF NEURALGIA.

I now approach what is really the most difficult portion of my task; for, although it would be easy enough to write copiously on the treatment of neuralgia, it is extremely difficult to keep a just medium between the opposite extremes of undue meagreness and of useless profusion of detail in the handling of this subject. There are also difficulties connected with the present uncertain and transitional state of opinion, even among high authorities, as to the value of particular remedies, and even of large groups of remedial agents, altogether there has been more hesitation in my mind as to this part of the present work than about any other, and the present chapter has been rewritten more than once. I mention this only to account for what there may very likely be found in it—an imperfect literary style such as too commonly marks work which has been repeatedly patched and corrected. At the same time, it should be said that my hesitation does not apply to the main principles of treatment which will be recommended below; it proceeds rather from the fear of seeming to ignore from carelessness modes of treatment which are still much used, but which I have really rejected, because, after full trial, they appeared to me valueless. Space is, after all, limited, and a complete account of all the remedies for neuralgia in vogue, in English and Continental clinics, would of itself fill a large volume.

The treatment of neuralgia may be divided into four branches: (1) Constitutional remedies; (2) narcotic-stimulant remedies; (3) local applications; (4) prophplaxis.

1. Constitutional treatment must be subdivided, as (*a*) dietetic, (*b*) anti-toxic, and (*c*) medicinal tonic.

(*a*) The importance of a greatly-improved diet for neuralgic patients is a matter which is more fully appreciated by the English school of medicine than by either the French or the German; it has, for instance, very much surprised me to notice the almost entire silence of Eulenburg on this topic. For my part, the opinions expressed three years ago* on this matter have only been modified in the direction of increasing

---

*Art. "Neuralgia" ("Reynold's System of Medicine," vol. ii. 1868.)

certainty; I have learned by further experience that the principle is even more extensively applicable than I had supposed.

That neuralgic patients require and are greatly benefited by a nutrition considerably richer than that which is needed by healthy persons, is a fact which corresponds with what may be observed respecting the chronic neuroses in general; and it gives me much satisfaction to point out this position of neuralgia as belonging to this large class of disorders, not merely by its pathological affinities, but by its nutritive demands. In a very excellent and suggestive paper by Dr. Blandford* it is stated, as the result of a large experience in mental and other nervous disorders, that the greater number of chronic insane and hypochondrical cases, as well as neuralgic patients, are remarkably benefited by what might seem at first sight almost a dangerously copious diet. Occasionally it happens that the patients discover this by the teaching of their own sensations, and the apparent excesses in eating which some epileptic and hypochondriacal persons habitually commit are looked on by many practitioners as the mere indications of a morbid *bulimia* which represents no real want, but only the craving of a perverted sensation which ought to be interfered with and allayed rather than encouraged. It is now many years since I began to doubt the justice of this opinion; the particular instance which called my attention to it being that of epilepsy, of which disease I saw a considerable number of cases, within a short period of time. that were distinguished by the presence of enormous appetite for food; and I finally came to the conclusion that, so far from this symptom being of evil augury, and likely to lead to mischief, it is, with certain limitations, a most fortunate occurrence. It is hardly necessary to say that overeating, such as produces dyspepsia and distention of a torpid intestine with masses of fæces, may distinctly aggravate the convulsive tendency; but the truth is that, with a little careful direction and management of the unusual appetite, these bulimic patients can in most cases be allowed to satisfy their desires without harm of this kind following; a larger portion of food really gets applied to the nutritive needs of the body, and the nervous system unmistakably benefits thereby, the tendency to atactic disorder being visibly held in check.

That which I have thus observed in the case of epilepsy, and which Dr. Blandford more particularly affirms concerning chronic mental diseases and the large number of neuroses that hover on the verge of insanity, has been most distinctly verified in my experience of the treatment of neuralgia. It is, unfortunately, by no means a frequent occurrence that the sufferer from this malady is inclined to eat largely, but the few patients of this type that I have seen were, in my judg-

---

* Practitioner, vol. iv., 1870.

ment, distinctly the better for it. Far more common in neuralgia is a disposition of the patient to care little for food, to become nice and dainty, and in particular to develop an aversion—partly sensational and partly the result of morbid fear about indigestion—for special articles of diet. Dr. Radcliffe pointed out the special tendency of neuralgics to neglect all kinds of fat; partly from dislike, and partly because they believe it makes them "bilious;" and I have had many occasions to observe the correctness of this observation. In fact, by the time patients have become sufficiently ill with neuralgia to apply to a consulting physician, they have already, in the great majority of cases, got to reject all fatty foods, and have cut down their total nutriment to a very sufficient standard. Young ladies suffering from migraine are especially apt to mismanage themselves, to a lamentable extent, in this direction: this is natural enough, because the stomach disorder seems to them the origin of the pain, instead of being, as it is, a mere secondary consequence of the neurosis. But it is not only the sufferers from sick-headache in whom we find this tendency to insufficient eating, especially of fat; not to mention that all severe pain usually tends to disorder appetite and make it fastidious, there is nearly always some wiseacre of a friend at hand, ready to suggest that neuralgia is something very like gout, that gout is always aggravated by good living, and, *ergo*, that the patient should be "extremely cautious as to diet;" the end of which is that the poor wretch becomes a half-starved valetudinarian, but, so far from his pain getting better, it steadily becomes worse. I cannot too strongly express the benefits that I have seen accrue, in the most various kinds of neuralgic cases, from persistent efforts to remedy this state of things, and to convert the patient from a valetudinarian to a hearty eater; and I wish particularly to say that this success has always been most marked when I have from the first insisted on fat forming a considerable element of the food. Cod-liver oil is the form in which I much prefer to give it, if this be possible; there can be no mistake about the relatively greater power of this than of any other fatty matter, I believe simply from its great assimilability. But the very cases in which we most urgently desire to give fat are often those in which the patient's fantastic stomach openly revolts at the idea of the oil; we must then try other fats; and we should go on trying one thing after another—butter, plain cream, Devonshire cream, even olive or cocoanut oil (though these are the poorest things of the sort we can use)—till we get the patient well into the way of taking a considerable, if possible a decidedly large, daily allowance of fat, without provoking dyspepsia. It is surprising what can be done in this way by perseverance and tact, and it is no less striking to observe the good effects of the treatment. Nothing is more singular than to see a girl, who was a peevish, fanciful, and really very suffering migraineuse, brought to a state in which she will eat spoonful after spoonful of Devonshire cream, and at the same time lose her headaches, lose her sickness, and develop the appetite of a day-laborer; and, though such very marked instances as this are uncommon, they do sometimes occur, and a minor but still important degree of improvement is very frequent.

As for the *modus operandi* of the fatty food, there is no certainty. Dr. Radcliffe believe it acts as a direct nutrient of the nervous centres; and I also cannot help feeling that there is some evidence in favor of this idea. But, whether this be so or not, there is another kind of action of fat that is more simple and obvious; namely, it seems to be certain that the enrichment of the diet by fat greatly assists the assimilation of food in general, and thus the patient's nutrition is altogether improved.

It is not merely, however, by increasing any one element of food that we should seek to enrich the diet of neuralgics, but rather by such a steady and persistent effort as Dr. Blandford describes, to increase the total quantity of nutriment to perhaps as much as one-third more than the patient would probably have taken in health. To those who from prejudice are incredulous of the propriety of this method, I would say, "Try it, and I venture to say your incredulity will disappear." More especially I would urge the great importance of this system in modifying the nervous status of very young, and also of aged, sufferers from neuralgia; it is the indispensable basis of a sound treatment for such patients.

This seems the proper place for such remarks as must be made upon the function of alcohol in neuralgia; for, though this agent is a true narcotic when given in large doses, it is not under that aspect that I can recommend its use in neuralgia at all. I have written so much on this subject lately, that I shall here content myself with an emphatic repetition of my protest against the use of alcoholic liquors as direct remedies for pain. They ought only to be given, in neuralgia, in such moderate doses, with the meals, as may assist primary digestion without inducing any torpor, or flushing of the face, or artificial exhilaration. I cannot too expressly reprobate the practice of encouraging neuralgics, especially women, to relieve pain and depression by the direct agency of wine or spirit; it is a system fraught with dangers of the gravest kind.

(*b*) The anti-toxic remedies include agents addressed to the modification of a special condition of the blood and tissues induced by the presence of morbid poisons, of which syphilis, malaria, and (more doubtfully) gout and rheumatism, are the representative examples.

Of syphilitic neuralgia the treatment may be summed up in a few words: Give iodide of potassium in doses rapidly increased up to a daily quantum of twenty to thirty grains. If this fails, give one-twelfth of a grain of bichloride of mercury thrice daily.

Of malarial neuralgia I can only speak from such a limited experience that I am by no means in a position to give an exhaustive account of the treatment. Quinine is, of course, the remedy that should first be tried; and, as the paroxysms are usually regular in their recurrence, I prefer to give the drug after the plan which is, I think, incontestably the best in ordinary ague—*i. e.*, to administer one large dose (five to twenty grains) about an hour before the time when the attack is expected. With a few exceptions the malady, unless it had taken very deep root before we were consulted, will yield to a few doses given in this way; after the morbid sequence has been thus interrupted, it will be proper to continue the action of quinine in smaller and more frequent doses, given for three or four weeks continuously. For the comparatively rare cases in which quinine fails, the prolonged use of arsenic (Fowler's solution, five to eight minims three times a day), especially with the simultaneous employment of codliver oil, is to be recommended.

The part which gout may play in inducing neuralgia is, as I have already said, a far more doubtful question than the popular medical traditions assume it to be; and treatment directed to gout as a cause is an extremely uncertain affair. The direct relief of neuralgic pain by the administration of colchicum, for example, is, in my experience, a very rare occurrence, even where the gouty diathesis is unmistakably present; and, on the other hand, the depressed vitality which gouty neuralgics usually show in a marked degree, renders it very doubtful whether the relief of the pain may not be too dearly purchased at the cost of the general lowering effects of colchicum. It is probable that neuralgia occurring in gouty subjects is more safely, and equally effectually, treated upon general principles. At the same time it may be admitted that, in the subordinate function of an adjuvant to the aperients which it is sometimes advisable to give, small doses of the acetic extract of colchicum seem to possess some value.

The question of treatment addressed to a supposed rheumatic element in neuralgia will, of course, be differently judged according to the respective ideas of various practitioners as to the pathological affinities of the two diseases; and the reader already knows that I believe these affinities to be different in kind from what is generally believed. The utmost that I should concede is, that in a certain very limited number of cases the peripheral factor in neuralgia is an inflammation of the nerve-sheath, or surrounding tissues, which forms part of a chain of phenomena of local fibrous inflammations in different parts of the body. Iodide of potassium, in five or ten grain doses three times a day, is the proper treatment for such cases. I have never found alkalies do any direct good to the pain.

(c) The medicinal tonic variety of constitutional treatment is more especially represented by the use of iron and arsenic in cases where poverty of the blood seems to exist in a marked degree, and by the administration of certain tonics—quinine, phosphorus, strychnia, and zinc—which are supposed to exert a specially restorative influence upon the nervous tissues.

The use of quinine as an anti-malarial agent has been already referred to; its employment in non-malarial cases is of much more restricted scope and benefit. Experience has taught me to agree in general with the opinion of Valleix, that it is a very unreliable agent; the one marked exception to this being the case of ophthalmic neuralgias. What the reason may be I cannot in the least say, but it is a fact that quinine does benefit these neuralgias, in cases where there is no room for suspicion of malaria, with a frequency which is very much greater than in the treatment of the painful affections of any other nerve in the body. The quantity given should be about two grains three times a day.

The preparations of phosphorus which I have employed in the treatment of neuralgia are the phosphuretted oil, the hypophosphite of soda (five to ten grains three times a day), and pills of phosphorus (according to Dr. Radcliffe's recommendation) containing one-thirtieth of a grain, given twice or thrice daily. Either of the two last will do all that phosphorus can do, but its utility is not very extensive or reliable. I have found it to do most good in cases where there was a high degree of anæsthetic complication.

Preparations of zinc have, in my hands, done no particular good, although I have tried them in all manner of doses.

Strychnia, on the other hand, is a remedy which I have learned to prize much more highly during the last few years than previously. Its most decided efficacy has been shown in some of the visceralgiæ, especially gastralgia, and (to a less extent) angina pectoris. Its internal use for these complaints is best effected by giving doses of five to ten minims of tincture of nux-vomica three times a day; but a method which I have several times employed with good effect is the subcutaneous injection of very small doses of strychnia (one-eighteenth to one-fiftieth of a grain) twice daily. For the superficial neuralgias, on the other hand, I generally administer one-fortieth of a grain, with ten or fifteen minims of tincture of sesquichloride of iron, by the stomach, three times a day; this is a very powerful prophylactic remedy to prevent the recurrence of the attacks when once the sequence of them has been broken through by other means.

Of iron generally, as a remedy in anæmic cases, I have only to remark that, in order to get its full benefits, it is necessary to use large doses. I give the saccharated carbonate in twenty-grain doses twice or three times a day.

But of the sesquichloride of iron I am inclined to say something more; it has seemed to me that, besides its effects on the blood, it has a marked and direct influence upon the nervous centres, which is different from anything which one observes in the action of other preparations of iron. It is certain that the action of sesquichloride of iron, in those cases of chlorosis which are distinguished by profound nervous depression, is something quite peculiar; and the effect which it produces in the anæmic neuralgias, more especially of young women, is equally remarkable. I cannot help alluding here to the striking effects which large doses of the tincture, as recommended by Dr. Reynolds, produce in acute rheumatism; the severest pain is often checked within twenty-four hours after the commencement of this treatment. Both in this disease and in neuralgia, I employ the old-fashioned tincture: if given alone it should be used in large doses (thirty or forty minims three times a day); but an excellent combination is that, already mentioned, of ten-minim doses of this tincture with one-fortieth of a grain of strychnia. There is something in the revivifying effects of this mixture that is quite peculiar. I have very lately employed it in the case of a gentleman, aged thirty-five, who was the subject of frontal neuralgia complicated with paralysis of the internal rectus, and who was decidedly anæmic, and greatly depressed and worried in mind by the consciousness of his inability to overtake professional work which had accumulated upon him. This patient improved with great rapidity, and in the course of three weeks lost, not merely his neuralgia, but also his strabismus, almost entirely; but he then got into a condition which, though not of permanent importance, was sufficiently undesirable to make me mention it here, especially as I have seen the same thing in more than one patient besides him. It is a peculiar state of restlessness during the day and sleeplessness at night, without any positive exaltation of reflex excitability such as one used to see from strychnia in the days when mischievously large doses of that drug were very commonly given, and patients used to complain of decided twitchings and startings of the limbs. It is clearly not a strychnia effect pure and simple, nor an iron effect only; it is a *tertium quid* compounded of the actions of both drugs.

The direct effects of arsenic in the improvement of the quality of the blood seem to me incontestable; and its use for this purpose in anæmic neuralgias is certainly something over and above its special neurotic action. No one, who has employed it much in the cases of anæmic children suffering from chorea after rheumatism, can have failed to observe its frequently striking influence upon blood-formation even long before the nervous ataxia is materially reduced. The misfortune is, however, that we possess no indications by which to judge beforehand whether we may reckon on its most favorable action in any given (non-malarious) case, with certain special exceptions. In angina pectoris it has a most direct effect, which is rarely altogether missed, and is sometimes surprising: the cases in which it succeeds best are those distinguished by anæmia, but we may well suppose, from its remarkable action upon other neuroses of the vagus, that it is something more than an action on the blood-making process which produces such powerful effects in allaying the tendency to recurrence of the paroxysms. My attention was called to its action in this disease chiefly by the remarkable case published by Philipp;* this was a purely neurotic angina, but one of the severest type, and the influence of arsenic was very striking. Since that time I have employed it in several cases, and, after trying various forms of administration, I conclude that nothing is better than Fowler's solution, in does of three minims (gradually increased, if the remedy be well tolerated, up to eight or ten) three times a day. Unfortunately, there are some neurotic patients who cannot bear arsenic, the irritability of their alimentary canal is such that the drug always provokes vomiting, or diarrhœa, or both; this was the case with one of my patients, in whose case I had allowed myself to hope for the very best results from arsenical treatment. But where the patient tolerates it—and usually he tolerates it extremely well—the prolonged use of arsenic seems really to root out the anginoid tendency, or at least to confine it to the more trivial and manageable manifestations. I believe that in at least three patients, I have so completely broken down a succession of cardiac neuralgic attacks as to substitute for them a mere remnant of a tendency to "tightness at the chest" after any severe bodily exertion or mental emotion. It might be a question, in cases where the stomach does not tolerate the ordinary administration of the agent, whether it would not be worth while to try the effect of subcutaneous injection (two to four minims of Fowler), or inhalation of the smoke of arsenical cigarettes. But, in truth, it is not certain that even in this case we escape the characteristic effects of the drug upon those persons who are abnormally sensitive to it.

A remarkable instance of the beneficial influence of arsenic occurred in the case of a woman, aged forty-six, the solitary example of severe angina in a female that I have ever seen. [It is by no means uncommon, however, to see the milder forms of cardiac neuralgia in women; the remarkable statistics of Forbes, quoted in Chapter I., must certainly have been taken exclusively from cases of the severest type of the disease.]

---

* Berlin. klin. Wochensch., 1865.

This was a hospital patient, who had always suffered much from hysteria, and from childhood had been liable to hemicranic headache; she had entered on the period of "change" at the time the attacks began, but menstruation, though irregular, still continued, and, in fact, did not cease till four years later, long after the anginal attacks had been subdued. The patient had been attacked for the first time at the end of a heavy day's washing; she dropped on the ground with the sudden agony and faintness, and thought she should "never come to life again." The paroxysms returned five times within the next month, though not always so severely as on the first occasion; but the poor woman lived in a constant state of terror. On the occasion of her second visit to me, she had a most severe attack in the waiting-room at the hospital: being called to her I found her very nearly pulseless, gasping, and with the kind of complexion which is so suggestive of approaching death. She was recovered by a large dose of ether. It was a rather uncommon feature in this case that the pain was only at and around the lower end of the sternum, except that occasionally it shot along the sixth intercostal space. The employment of Fowler's solution (in doses gradually mounting to twenty-one minims daily) for six months completely eradicated the anginal tendency; the proof that it was a real therapeutic effect was given by the result of an attempt to leave the medicine off at the end of eight weeks' treatment; the patient immediately began to suffer again. When she really left off, at the end of six months' treatment, she had had no tendency to heart-pang for more than a month, and, besides this, looked quite another creature in her improved vitality and vigor. Yet the menstrual troubles went on, and the function was not finally suppressed for a long time afterward.

I suspect, however, that the most frequent successes with arsenic will, after all, be made in the cases of more or less anœmic male patients who are attacked with the neurotic form of angina in the midst of a career (as is especially the case with some professional careers) that implies not merely incessant labor, but great anxiety of mind. The drug does little good, however, if not positive harm, in that form of angina pectoris minor which is not the result purely of these causes, but of these, or some of these, plus the morbid action of the alcoholic excess, to which the patient has fled in order to relieve mental harassment and the fatigue that comes from overwork, especially overwork at tasks that are not congenial to his natural disposition; there is usually in such cases a heightened irritability of the alimentary canal, which is almost sure to cause arsenic to disagree: the really useful treatment is quinine for the first few days, and then, when the stomach will bear it, cod-liver oil in increasing doses, up to a large daily amount given for a long time together.

On the whole, arsenic, from its singularly happy combination of powers as a blood-tonic, a special stimulant of the nervous system, and withal as a special opposer of the periodic tendency, must be regarded as one of the most powerful weapons in the physician's hands, and (although it seems to act best in the neuralgias of the vagus and of the fifth) there is a possibility of its proving the most effective remedy in almost any given case which may come before us.

2. The narcotic-stimulant treatment for neuralgia includes some of the most powerful remedies for the disease which we possess. These remedies have very different properties, but they all agree in this, that in small doses they appear restorative of nerve-function—in large doses depressors of the same.

Four very different types, at least, of narcotic-stimulant drugs are useful in neuralgia: (a) There is the opium type, by which pain is very directly antagonized, and, besides this, sleep is also directly favored. (b) There is the belladonna type, by which pain is also much relieved, though with far greater certainty in some regions than in others (e. g., much the most powerful effect is seen in cases of pelvic visceralgia), but sleep is by no means so certainly or directly produced as by opium. (c) There is the chloral type, which is almost purely hypnotic; it is represented almost solely by chloral itself, which is resembled by scarcely any other drug. (d) There is bromide of potassium, which stands alone for its powerful action on the cerebral vaso-motor nerves, and which is useful in neuralgia simply by its power to check psychical excitement directly (through the circulation) and indirectly (through the production of sleep).

(a) Opium and the remedies that resemble it are, for the treatment of neuralgia, fully represented by the hypodermic use of morphia, which is the only kind of opiate treatment that ought ever to be employed, save in very exceptional instances. The great reasons for the preference of the subcutaneous administration over the gastric are, the economy of the drug which it affects and the much smaller degree of disturbance of digestion which it causes. The hypodermic injection of morphia, if conducted on correct principles, enables us, when necessary, to repeat the dose a great number of times with but little loss of the effect, and consequently with a much smaller rate of progressive increase of the quantity required; and the absence of depressive action on digestion enables us to carry out simultaneously that plan of generous nutrition which has already been shown to be so important a part of treatment. Indeed, the case is hardly expressed with sufficient strength, when we say that hypodermic morphia is usually harmless to the digestive functions; for in a great number of instances it will be found actually to give an important stimulus both to appetite and digestion; and the patient, who without its aid could hardly be persuaded to take food at all, will not unfrequently eat a hearty meal within half an hour after the injection.

The remarkable effects of hypodermic morphia have, however, caused it to be rashly and indiscriminately used, and so much harm has been done in this way that it is necessary to be exceedingly careful in the rules which we lay down for its employment. Upon these grounds I must hope to be excused if, in order to render this work complete, I repeat a good deal of what I have already said in other places. In the first place, I shall speak of the mode of administration, and then of the dose.

As regards the mode of administration, I prefer the use of a solution of five grains of acetate of morphia to the drachm of distilled water; if the acetate be a good specimen, this will dissolve easily (and keep some time without precipitation) without the use of any other solvent. With a solution of this strength we require nothing elaborate in the form of the syringe; a simple piston arrangement does well; only it is advisable that the tube shall have a solid steel triangular point, and a lateral opening. As regards the place of injection, I must repeat the opinion[*] which I have already published, that Mr. Hunter's plan of injection at an indifferent spot is, in the great majority of instances, fully as effective as the local injection would be; nevertheless, there is one consideration which in some cases may properly induce us to adopt the latter plan. Very nervous and fanciful patients will sometimes be much more readily brought to allow the operation when it seems to go directly to the affected spot, when they would be sufficiently incredulous of the benefits of an injection performed at a distance to indulge their dislike of incurring pain by refusing to submit to it. And there is one class of cases in which it is likely that there are real physical advantages in the local injection; in instances of old-standing neuralgia with development of excessively tender "points," which are also the foci of the severest pain, it will sometimes be advisable to inject into the subcutaneous tissue at these points. There is undeniable reason for thinking that the sub-inflammatory thickening of tissues around a certain point of nerve delays the transit of the morphia into the general circulation (and enables it to act more directly and powerfully on the nerve, which it thus renders insensitive to external impressions; an important respite is thus gained, during which the nerve-centre has time to recover itself somewhat. At the same time it must be remarked that this immediate injection

---

[*] In a paper on the "Hypodermic Use of Remedies," in the *Practitioner* of July, 1868, I gave the reasons for this opinion in full; and I see no reason to alter any thing I then said.

of a tender point is apt to be exceedingly painful, and it may be absolutely necessary to apply ether-spray before using the syringe. In early stages of neuralgia, before the formation of distinct tender points, there is no advantage whatever (except the indirect one above mentioned) in the local injection. And, on the other hand, it is often of great consequence not to run the chance of disfiguring such a part as the face, the neck, etc., when the injection can easily be done over the deltoid, or in the leg, or in some other part which even in women is habitually covered by the dress.

The dose to be employed is an exceedingly important matter, and one as to which practitioners are still very often injudicious. We ought never to commence with a larger dose than one-sixth of a grain; but very often as little as one-twelfth of a grain will give effective relief, and in not very severe cases it is well worth while to try this smaller quantity. When no larger quantity than one-sixth of a grain is employed we commonly observe no narcotic effects, *i. e.*, there is no contraction of pupil, no heavy stupor, and, although the patient very often falls asleep, on waking he does not experience headache, nor is his tongue foul. I cannot too strongly express the opinion that it is advisable by all means to content ourselves with this degree of the action of hypodermic morphia, unless it fails to produce a decided impression on the pain. But in very severe cases our small doses will fail; and then, rather than allow the patient to continue having severe paroxysms unchecked, we must frankly admit the necessity of using a narcotic dose from one-quarter to one-half of a grain, according to circumstances. Whatever actual dose be employed, it is important not to repeat it with unnecessary frequency; once a day in the milder, and twice a day in the more severe cases, will be all that is advisable, save in very exceptional cases: the point being to administer it as quickly as possible after the commencement of an exacerbation. If by these means we can prevent the patient having any severe pains during a period of several days, we often give time to the affected nerve to recover itself so completely, especially with the aid of other measures to be presently mentioned, that the tendency to neuralgia is completely broken through, and we can drop the injections, either at once or by rapid diminution of the dose, and thereafter treat the case merely with tonics, and with the precautionary measures to be dwelt upon under the heading of Prophylaxis. But, if we have been driven to to the use of distinctly narcotic doses, and these do not very speedily break the chain of neuralgic recurrence, it will not do to continue to rely upon hypodermic morphia; it will be best to try some of the local remedies (blistering, galvanism) with it. If this combination fails, we should then try the effect of atropine, the sulphate of which, hypodermically injected, fully represents for all useful purposes the mydriatic class of narcotics.

(b) The commencing hypodermic dose of atropine should be one-one hundred and twentieth grain; it is not often that so small a quantity will do any good, but it is necessary to use this agent with great precaution, as we occasionally meet with subjects in whom extremely small doses provoke most uncomfortable symptoms of atropism, as dry throat, dilated pupil, delirium, and scarlet rash. Commonly we shall find ourselves obliged to increase the dose to one-sixtieth, one-fiftieth, or one-thirtieth of a grain; and in a very few cases it may be necessary to go even as high as the one-sixteenth or one-twelfth. In my experience such instances are excessively uncommon; and I cannot but suppose that the practitioners who use the high doses frequently must inject in such a manner as to fail to get the whole dose taken up. [Absolutely inexplicable to me is the statement of the illustrious Trousseau—that hypodermic remedies are "less active" (!) than gastric remedies—except on his hypothesis.]

The most remarkable effects that I have seen from hypodermic atropia were obtained in cases of peri-uterine neuralgia, especially dysmenorrhœal neuralgia. Speaking generally of atropine, it must undoubtedly be counted far inferior to morphia as a speedy and reliable reliever of neuralgic pain, but for all pelvic neuralgias it appears to me on the whole to surpass morphia. And besides this, in other neuralgias, where opiates altogether disagree (as with some subjects they do), it is not uncommon to find that atropia acts with exceptionally good effect. And to some extent I am inclined to confirm Mr. Hunter's opinion, that, where atropia does stop neuralgia, it does so more permanently than morphia.

There is another special use of hypodermic atropine which I have not seen mentioned by any one but myself, but which is probably very important, namely, in ophthalmic neuralgia where acute iritis, or especially glaucoma, seems coming on. I may be mistaken, but I believe that in three cases I have succeeded, by prompt injection of sulphate of atropine (one-sixtieth to one-fortieth of a grain), in saving a neuralgic eye from damage. and possibly from destruction, from impending glaucoma.

(c) The class of cases for which merely hypnotic remedies are of much value is limited; nevertheless, in the milder kinds of migraine and clavus, especially when they have been brought on or are kept up by mental worry or hysterical excitement, these remedies will sometimes prove very useful. In former days, before we knew chloral, I used to employ camphor for this purpose; three or four grains being administered every two hours: and in hysterical hemicrania of a not very severe type this not unfrequently produced a short sleep, from which the patient awoke free from the pain. But chloral infinitely transcends in value any agent of this kind that was known before. Perfectly valueless for the really severe neuralgias, it is of the greatest possible use as a palliative in migraine and clavus, where the great object, for the moment, is to get the patient to sleep. A single dose of twenty to thirty grains will often effect our object: it may be repeated in two hours if sleep has not been induced; it should be given as soon as the pain has at all decidedly commenced.

And here I wish to make some special remarks on the subject of "palliation," and the relation it bears to "cure." Nothing is more common than to read serious admonitions, in medical works, about the folly of trusting to remedies which only palliate for the moment. but leave the root of evil untouched; and, of course, there is a certain respectable modicum of the fire of truth behind all this orthodox smoke. In the case of neuralgia, however, it is most important to understand that mere palliation, that is, stopping of the pain for the moment, may be either most useful or highly injurious, according to the way in which it is done. The unnecessary induction of narcosis for such a purpose, doubtless, is most reprehensible; but if it were possible simply to produce sleep from which the patient should awake refreshed, without any narcotic effects, then, certainly, that sort of palliation must be good. That is precisely what the judicious use of chloral does; and I may mention, as resembling though not equalling it, the action of Indian hemp, which has been particularly recommended by Dr. Reynolds. From one-fourth to one-half of a grain of good extract of cannabis, repeated in two hours if it has not produced sleep, is an excellent remedy in migraine of the young. It is very important, in this disease, that the habit of long neuralgic paroxysms should not be set up; and if the first two or three attacks are promptly stopped, by the induction of sound, non-narcotic sleep, we may get time so to modify the constitution, by tonics and general regimen and diet, as to eradicate the neuralgic disposition, or at least reduce it to a minimum. But I would decidedly express the opinion that such remedies as either opium or belladonna are mostly unsuited to this purpose. If the migraine of young persons does not yield to chloral, to cannabis, or to muriate of ammonia (in twenty or thirty grain doses), it will not be advisable to ply the patient with any remedies of the narcotic-stimulant class, but to trust to tonic regimen and the use of galvanism.

The mention of muriate of ammonia, which, for migraine and clavus and the milder forms of sciatica, not unfrequently proves useful in stopping the violence of a paroxysm and enabling the patient to get some refreshing sleep, leads me to noeice that not only may a variety of the milder narcotic-stimulants be employed in this way, but the external stimulus of heat to the extremities (very hot pediluvia) greatly assists the action of any such remedies; especially if mustard-flour be added, so that a mild vapor of mustard rises with the steam and is inhaled. Perhaps the ideal medication, to arrest a bad sick-headache, is to give twenty grains of chloral, and make the patient plunge his feet in very hot mustard-and-water and breathe the steam. He can hardly fail to fall asleep for a longer or shorter time, and awake free from pain.

(*d*) The use of bromide of potassium in neuralgia is a subject of great importance, and which requires much attention and discrimination. In common with, I dare say, many others, I made extensive trial of this agent when it first began to be much talked of, but was so much disappointed with its effects in neuralgias, that at one time I quite discarded it in the treatment of those affections. Renewed experience has taught me however, that, though its use is restricted, it is extremely effective if given in appropriate cases and in the right manner. For the great majority of neuralgias it is quite useless, and, what is more, proves often so depressing as indirectly to aggravate the susceptibility of the nervous system to pain. The conditions, *sine quis non*, of its effective employment seem to be the following: The general nervous power, as shown by activity of intelligence, and capacity of muscular exertion and the effective performance of co-ordinated movements, must be fairly good, find the circulation must be of at least average vigor; the patient must not have entered on the period of tissue-degeneration. Among neuralgics who answer to this description, those who will benefit by the bromide are chiefly subjects—especially women—in whom a certain restless hyperactivity of mind and perhaps of body also, seems to be the expression of Nature's unconscious resentment of the neglect of sexual functions. That unhappy class, the young men and young women of high principle and high mental culture to whom marriage is denied by Fate till long after the natural period for it, are especial sufferers in this way and for them the bromide appears to me a remedy of almost unique power. But I wish it to be clearly understood that it is not to the sufferers from the effects of masturbation that I think the remedy specially applicable: on the contrary, it is rather to those who have kept themselves free from this vice, at the expense of a perpetual and almost fierce activity of mind and muscle. The effects of solitary vice are a trite and vulgar story; there is something far more difficult to understand and at the same time far more worth understanding in the unconscious struggles of the organism of a pure minded person with the tyranny of a powerful and unsatisfied sexual system. It is in such cases, which it needs all the physician's tact to appreciate, that it is sometimes possible to do striking service with bromide of potassium; but it will be necessary to accompany the treatment with strict orders as to generous diet, and, very likely, with the administration of cod-liver oil.

Having decided that bromide of potassium is the proper remedy, we must use it in sufficient doses. Not even epilepsy itself requires more decidedly that bromide, to be useful, shall be given in large doses. It is right to commence with moderate ones (ten to fifteen grains), because we can never tell, beforehand, that our patient is not one of those peculiar subjects in whom that very disagreeable phenomenon—bromic acne—will follow the use of large doses. But we must not expect good results till we reach something like ninety grains daily. let me add that it is not so far as I know, by reducing any "hyperæsthesia" of the external genitals, of which the patient is aware, that the remedy acts; I have not seen such a nexus of disease and remedy in these cases.

3. *Local Measures.*—The external remedies which may be applied for the treatment of neuralgia may be divided into (*a*) skin-stimulants ; (*b*) paralyzers of peripheral sensory nerves: (*c*) remedies adapted to diminish local congestion; (*d*) remedies adapted to diminish arterial pulsation; (*e*) electricity; (*f*) mechanical means of protection.

(*a*) Among the skin-stimulants blisters hold the highest place as a remedy for neuralgia; indeed the assertion of Valleix, that they are the best of all remedies, is still not very wide of the truth. They are by no means universally applicable, and the degree to which their action should be carried varies materially in different forms of the disease, but they are of the greatest possible service in a large number of instances.

It is possible to view the action of blisters in neuralgia in more than one way. When applied in such a manner as to vesicate decidedly, and especially if kept open and suppurating for some time, they cause considerable pain of a different kind from that of neuralgia itself and the mental effect of this, operating as a diversion of the patient's thoughts from his original trouble, may be thought to assist in breaking the chain of nervous actions by which he is made to feel neuralgic pain. There may be something in this, but I confess that I do not believe this kind of effect goes for much in genuine neuralgia. It is rather in the pain of hypochondriasis, and the so-called spinal irritation (to be described in the second part of this work), that such an action of blisters proves useful.

Another action of blisters, which some authors hold to be perhaps the most effective portions of their agency, is that which is produced by the drain of fluid, specially when they are kept open, by which means a kind of depletion is set up, and the morbid irritation that causes the nerve pain removed. I cannot at all assent to this view. In the first place, I believe that any one who has large experience of blistering in neuralgia will ultimately come, as Valleix did, to believe that prolonged drain from a blister is rarely or never useful, and that a far better plan is that of so-called flying blisters, renewed at intervals if necessary. The most genuine successes that I have procured from blistering have certainly been got in this way. But I should go further, and say that the prolonged drain and the peculiar kind of chronic irritation produced by a suppurating blistered surface can very decidedly aggravate a neuralgia; this is more especially the case when the blister is applied immediately over the focus pain.

The view which I am strongly convinced alone explains the beneficial action of blisters is that which supposes them to act as true stimulants of nerve-function. In order that this effect shall be produced, it will be necessary that the skin-irritation be either produced at some distance from the seat of the greatest pain, or that, if applied in that spot, it shall be comparatively mild in degree. And accordingly, I have been led, in my observations to apply the blister at some distance from the focus of pain. An indifferent point, however, will not do—there must be an intelligible channel of nervous communication between the irritated portion of skin and the painful nerve. This object is accomplished by placing the blister as close as may be to the intervertebral foramen from which the painful nerve issues; the effect of this is probably a stimulation of the superficial posterior branches, which is carried inward to the central nucleus of the nerve. I must say that the results which I have derived from this plan of treatment have been far more satisfactory than those which I used to obtain when I habitually applied the vesication as near as might be to the focus of peripheral pain; and I think that this result tallies well with the idea that the essential mischief in neuralgia consists in an enfeebled vitality of the central end of the posterior root. An exceedingly interesting confirmation of this idea as to its *modus operandi* has been afforded me by the fact that not merely neuralgic pain, but also trophic and inflammatory complications attending it, have been sensibly relieved, in several cases that I have seen, by this mode of reflex stimulation. This has been particularly the case in herpes zoster, where the process of inflammation and vesiculation has been very promptly checked by the application of a tolerably powerful blister by the side of the spine at the proper level; and I am gratified to mention that Dr. J. K. Spender, of Bath, pointed out this fact* at a time when he had only seen my statement that the pain could be relieved in this way. In the case of the trigeminus, the same kind of reflex stimulation is most effectively obtained by applying the blister over the branches of the cervico-occipital, at the nape of the neck; and it is remarkable what powerful effects are sometimes thus produced, even in cases that wear the most unpromising aspect. For example, in the desperate epileptiform tic of old age, I have more than once seen a complete cessation of suffering, which lasted for a very long time—so long, in fact, as to make me hope against hope that it might never return. I do not now entertain any such expectations from this remedy; still, its value is very great.

There are curious differences between the effects of blistering in trigeminal or intercostal neuralgia and in sciatica. On the whole, it would appear that blistering in the neighborhood of the spine is less frequently effective in the latter, and we sometimes, after failing with this method, obtain immediate success by two or three repetitions of the flying blister, somewhere over the trunk of the nerve, especially just outside the sciatic notch. I have one lady patient in whom this series of phenomena has several times been observed; and I have seen it occur in a particular attack, in other patients, in whom, nevertheless, on another occasion the spinal blistering has been promptly effective.

I consider blistering of the posterior branches to be an important, and usually an essential, element in the treatment of all cases of sciatica in the middle period of life which have reached some severity and lasted long enough to become complicated with decided secondary affections.

In all cases where blistering is employed it is advisable to adopt the simultaneous use of hypodermic morphia or atropine; this combination of remedies is exceedingly powerful.

Lastly, it must be said of blistering, that, on the whole, it is a remedy not well fitted to be applied to aged subjects; and in its severer forms it should never be applied to patients who are greatly prostrated in strength. For it must be borne in mind that the remedy may miss its aim of relieving the neuralgia, in which case it is necessary to remember, more accurately than many practitioners appear to do, what a very serious element of misery and prostration will be introduced into the case by the vesication itself.

I am not convinced that any of the other forms of severe skin-irritation (*e, g.*, tartar-emetic inunction, or the use of veratrine-ointment to such a degree as to produce not the anæsthetic but the irritant effects) are of any particular value; if blistering failed, I should not expect to see them succeed.

A milder degree of skin-stimulation is represented by rubefacient liniments of various kinds, which may be briskly rubbed into the skin along the track of the painful nerve, without any danger of producing vesication. Among this class I continue to prefer chloroform diluted, with six or seven parts of chloroform, to any other; in the milder forms of neuralgia, especially in young persons and first attacks, it is surprising how frequently tne paroxysm may be greatly relieved, if not arrested. Still, this can only be regarded as the merest palliative; and in severer cases such applications are useless. Occasionally, when chloroform-liniment has failed, a mustard plaster will do good.

The mildest degree of skin-stimulation is represented by the continuous application of moist warmth, which is best effected by the simple application of moistened spongio-piline; so far as I have observed, however, it is rather in cases of myalgia than in true neuralgia that this does good; in the latter it is probably little more than a mere protector against cold.

(b) A variety of agents can be employed with the object of temporarily interrupting the conductivity of the painful nerve; by this means a period of rest is obtained during which the centres,—sensory and psychial—have time to regain a juster equilibrium, and the habit of pain is, *pro tanto*, broken through.

There is one agent of this class which for general purposes I do not think is worth retaining on our list of sensory paralyses —namely, cold. Cold, to be of any value, ought to be of the degree which is represented by ice allowed to melt slowly in contact with the skin; and for the majority of neuralgias this is decidedly inferior to other remedies that can be applied by painting or inunction. The one case in which ice is supremely useful is in neuralgia of the testis; here I make no doubt that it is almost, if not quite, the most useful remedy we can employ, although of course other means must be taken to modify the neuralgic temperament. It should be applied the moment an attack comes on.

Far more useful, in neuralgias generally, is the external application of aconite or of veratrine. Aconite may be employed in the milder or the stronger form; in the former case, we simply paint the ordinary tincture on the skin over the painful nerves (avoiding any cracks or sores); in the latter, we rub in an ointment containing one grain of the best hydrate of aconitine to the drachm of lard, about twice a day, and to such an extent as to maintain complete numbness of the parts continuously, for two, three, or four days I do not believe that this will ever, by itself, cure a true neuralgia of any considerable severity; but I have more than once known its intervention, at a crisis in treatment when it seemed that other remedies might fail, produce a striking change in the progress even of a very bad case.

A milder, but still very useful form of the same kind of action, is produced by veratrine-ointment. I would recommend, however, as a rule, that it be employed, at any rate at first, of weaker strength than that recommended in the Pharmacopœia, for with some persons it is easy to pass the anæsthetic, and to enter on the irritant, action of veratrine upon the skin. This leads me to give a caution that should properly have come earlier, when I was speaking of skin-stimulants. In aged subjects, especially, we rather frequently meet, in neuralgia, with a specially irritable state of the skin, even although there may be at the same time some loss of common and tactile sensation; and the practitioner must be warned against the danger of producing an amount of skin-irritation which will fearfully annoy his patient. I speak feelingly, having by such an indiscretion lost the richest patient who ever favored my consulting-room with his presence!

The inunction of mild veratrine-ointment is extremely useful, as an adjunct to other treatment, in migraine and supra-orbital neuralgias of suckling women, and of chlorotic girls. I have also seen it do much good in mammary neuralgia.

The last division of the subject of paralyzing agents in the treatment of neuralgia includes the surgical operations for division or resection of a painful nerve. Upon this question there is much difficulty in speaking decidedly. I admit at once, of course, that surgical interference is evidently indicated when, along with decided and intractable neuralgic pain, there is plain evidence either of the existence of a neuromatous tumor, or the presence of a foreign body impacted, or a tight cicatrix pressing upon a nerve. I admit, also, though with much greater qualifications, that carious teeth may need to be extracted before we can cure a neuralgia; but even here I should put in the decided caveat that we must consider whether the system is in a state to bear the shock, and that in any case we probably ought to mitigate the effects of the operation by performing it under chloroform. And I need hardly tell any one, who is familiar, either practically or from reading, with the subject, that thousands of carious teeth have been extracted from the mouths of neuralgic patients, not only without benefit, but with the effect of distinctly aggravating the disease. And I am yet more doubtful as to the advisability of such surgical procedures as the division or the resection of a piece of the painful nerve. Theoretically, as the reader will understand from the strong opinion I have given as to the mainly central origin of neuralgias, I never could anticipate that such a procedure would be more than temporarily successful; on the contrary, the mischief in the central end of the nerve remaining, I should suppose that the trying process of the reunion of the nerve (which always takes place) would be almost certainly attended with a revival of the neuralgia, too probably in an aggravated form. The only two cases of excision of a piece of the nerve, that I have ever seen, completely answered to this anticipation. In common fairness, however, I must admit that there is a large amount of evidence on the other side. Neuralgias of the trigeminus are pretty nearly the only cases in which the proposal of neurotomy or neurectomy ought to be entertained; in mixed nerves the inconvenience of the muscular paralyses that would follow would be usually too serious to allow of our incurring them. But resection of painful branches of the trigeminus has been performed in a great number of instances, more especially by German surgeons, with results that merit our attention; the cases recorded by Nussbaum, Wagner, Bruns, and Podratzki, may be especially referred to. On the other hand, with the exception of simple division of the nerve, which can be subcutaneously performed, and is a trivial proceeding (but has very short-lived effects), these operations are by no means without danger, especially when they are pushed to such a length as the opening of bony canals, and the resection of considerable portions of bone in order to get sufficiently far toward the centre, and fatal results have in more than one case followed. Above all, we can never too seriously reflect on the most interesting case of Niemeyer's reported by Wiesner,* in which the most formidable operations of this kind have been performed, in an apparently desperate case of epileptiform facial tic, and in which, after all, the application of the constant current painlessly effected an infinitely greater amount of good than had been done by all those severe and painful surgical manipulations. I think it is impossible, after this, not to conclude that neurectomy ought never to be even thought of except as a last resort, in cases of extreme severity, after other measures had been patiently tried and had decisively failed.

(c) Of remedies that are intended to relieve local congestion, I must speak with very doubtful approbation. Leeches or scarifications are, I think, very seldom of value. The only remedy that has sometimes seemed to do good is local compression, and, after all, it is quite as likely that this acts by anæsthetizing the nerve as by reducing congestion.

(d) Remedies that intefere mechanically with arterial pulsation are of considerable value where they can be effectively applied. I have already pointed out the specially aggravating effect of the momentarily-repeated shocks of arterial pulsation upon neuralgic pain. Where, then, it is possible, effectively to control an artery pretty near to the point where it divides into the branches that lie close to the painful part of the nerve, it is always worth while to try the experiment. But such a measure as the compression of the carotid in trigeminal neuralgia is of very doubtful propriety; I suspect the consequent anæmiation of the brain more than does away with any benefit that might be mechanically produced. And any attempt to intefere with the general arterial circulation by cardiac depressants is not to be permitted for an instant.

---

* Berlin. klin. Wochensch., 17, 1868.

(e) We enter now upon a most important subject, the treatment of neuralgia by electricity. It is necessrry to exercise much caution in speaking upon this topic, and, as I shall have to express somewhat decided opinions, I may be excused for referring to the circumstances under which I have arrived at my present stand-point upon this question. I can hardly be accused of having, with any very rash haste, espoused the cause of medical electricity in the therapeutics of pain, as any one will see who cares to turn to my article on Neuralgia* written only three years ago. At that time I had already been studying the subject for a considerable period, but was so convinced of the multitude of opportunities for fallacy that beset the student of electro-therapeutics, that I was unwilling to state more than the minimum of what I hoped and believed might be affected by this mode of treatment. Since that time I have become more fully acquainted with the researches of foreign observers, and, with the help of their indications, have been able to apply myself more fruitfully to my personal inquiries into the matter. The result is, that I am now able to speak with far greater assurance of the positive value of electricity as a remedy for neuralgic pain. I shall make bold to say that nothing but the general ignorance of the facts can account for the extraordinary supineness of the mass of English practicioners with regard to this question.

In the first place, I have arrived at a decided conviction that Faradic electricity is of little or no value in true neuralgias, and that the cases which are apparently much benefited by it will invariably be found, on more careful investigation, to belong to some other category.

On the effect of frictional electricity I have had such very small experience that I cannot venture to speak with any confidence, and the accounts that I have heard from others whose experience is much larger have not led me to attribute much importance to this agent. If I am to judge at all, I should say it merely acts as a skin-stimulant, and is, in that capacity, inferior to many other simpler and more facile applications.

Very different is the verdict of experience as regards the effects of the constant current; here the results which I have obtained have been so remarkable that even now I should distrust their accuracy, were it not that they are in accord with the general result which (among minor discrepancies) may be gathered, we may fairly say, from all the more important researches that have lately been carried out in Germany. The constant current, as I now estimate it, is a remedy for neuralgia unapproached in power by any other, save only blistering and hypodermic morphia, and even the latter is often sur-

---

* "System of Medicine," vol. ii.

passed by it in permanence of affect; while it is also applicable in not a few cases where blistering would be useless or worse.

The English medical profession has not as yet adequately appreciated the necessity for great care in the choice of apparatus and the mode of application of electricity. It is all-important, however, and especially in the case of applying galvanism for the relief of pain. The first quality that must be absolutely required in a battery, that is to be used for this purpose, is that it shall deliver its current with as little as possible variation of tension, in fact that it shall be constant, and not merely continuous; a vast majority of all the various galvanic apparatus that have been used have been merely the latter, and have consequently been almost valueless for the relief of pain. Such are Pulvermacher's chains, the voltaic piles made with elements of metallic gauze, Cruickshank's battery, and many others that have been used. A sufficiently constant current may be obtained from either of the following apparatuses. (1) Daniell's battery, (2) Bunsen's, (4) Smee's, For hospital use, the Daniell battery (in Muirhead's modification, or with the form of cells introduced by Siemen's-Halske) is perhaps the most desirable; but for-private practice it is worth while to sacrifice something of the superior constancy which we gain in the Daniell battery for the sake of comparative portability. All purposes which we aim at in the electric treatment of neuralgia may be sufficiently obtained by the use either of the Bunsen battery (zinc-carbon, excited by dilute sulphuric acid), as modified by Stohrer, or by the Smee battery (zinc and platinized silver, excited by dilute sulphuric acid), as in the highly convenient apparatus devised by Mr. Foveaux, of Weiss & Son's. It must be remarked that, for the purpose of treating neuralgia, we shall never need to employ more than fifteen, or at the utmost twenty, cells of either of these batteries. Both the Stohrer's Bunsen and the modified Smee of Weiss are made so that the elements are not immersed in the exciting fluid until the moment when the battery is going to be used; a simple mechanism at once throws the battery into or out of gear. In this way, destruction of the elements is minimized; and either of these two batteries may be used for from three to six months without any renewal, supposing the average work done to be one or two daily seances. If the battery is worked harder, it will require more frequent revivification. I strongly recommend London practitioners to deliver themselves from all care and trouble about the repair of their batteries, by making an agreement with the manufacturers to inspect and set them in order at stated intervals. The country practitioner, on the other hand, will do well to familiarize himself with the process of renewing the acid, of cleaning the plates, of amalgamating the zinc, etc.; in fact, to make himself independent of the manufacturer in every thing short of an actual renewal of the elements, when that becomes necessary. For all further details respecting the above-named, and other batteries, I must refer the reader to systematic works on medical electricity.* I must now pass on to the various modes of application, and the cautions to be observed.

It is, in the first place, necessary to say, that all the best observers coincide in the statement that the use of a current intense enough to produce actual pain or severe discomfort is never to be thought of in the treatment of true neuralgias; such practice will infallibly do harm. Only such a current is to be employed as produces merely a slight tingling, and (on prolonged application) a slight burning sensation, with a little reddening of the skin at the negative electrode. This being the case, it is perhaps not unnatural for those who have not had practical experience, to suspect that an application which causes so little palpable perturbation is devoid of any positive influence at all. Such skepticism will certainly not survive any tolerably lengthened observation of the actual facts; but, as some persons may be deterred by this *prima-facie* view of the case from making any fair trial of the current, it may be worth while, here, to allude to the unmistakable physical effects which similarly painless constant currents are repeatedly observed to produce in cases of motor-paralysis attended with a wasted condition of muscles. Those who have had experience of the treatment of such cases know that it is a by no means infrequent thing to see both muscles and nerves aroused from a state of complete torpidity, and brought into a condition in which the Faradic current, quite powerless before, is again able to excite powerful contractions, while, at the same time, the bulk of the muscles has increased most sensibly. These, surely, are sufficient indications of a positive action of the painless constant current; and such facts have now been recorded, in multitudes, by most competent observers.

The next maxim of first-rate importance is that the applications of the current should be made at regular intervals, and at least once daily; in most instances, this is enough, but occasionally it will be found useful to operate twice in the day. The matter of regularity is, I find, of great consequence, and it will not do to intermit the galvanism immediately on the occurrence of a break in the neuralgic attacks: it should be continued for some days longer.

The length of sittings is a point as to which there is considerable difference of opinion between various authorities; but

---

* The English reader may consult Althaus ("A Treatise on Medical Electricity," second edition, Longmans), or Meyer ("Medical Electricity," translated by Hammond: Trubner & Co.)

my own experience coincides with that of Eulenburg, that form five to ten, or, at the utmost, fifteen minutes, is almost the range of time.

Closely connected with the question of the length of sittings, is that of the continuity with which the current is to be applied. I have seen the best results, on the whole, from passing a weak current, wtthout any breaks, for about five minutes. But, where there are several foci of intense pain, it will often be advisable to apply the current to each of these, successively, for three or four minutes.

The places to which the electrodes should be applied vary much according to the nature of the case.

Benedikt's rule, that the application of electricity, to be useful, must be made to the seat of the disease, is undoubtedly true; but it is capable of being applied in a somewhat different manner from that which he recommends in particular cases, the difference being due to the view of the pathology of neuralgia which is taken in this work. That view is, that the essential *locus morbi* is always in the posterior nerve-root (and usually in that portion of the root which is within the substance of the cord), and that the peripheral source of irritation, if any, is only of secondary—though sometimes of considerable—importance. Hence the main object, in electrization, would seem to be to direct the influence of the current upon the posterior nerve-root. This may, however, be done in different ways, according to the situations in which we place the electrodes, and the direction in which we send the current.

There are, as yet, very considerable differences of opinion among electro-therapeutists as to the principles which should govern us, both in the localization of the effect and the direction of the current. Benedikt, for example, recommends that the current should be directed toward the supposed seat of the mischief. Thus, if we suppose a neuralgia to depend on morbid action within the spinal cord, then we may galvanize the spine, taking care to make the current come out through any vertebra over which we detect tenderness. If we suppose the seat of the disease to be in the nerve-root in the mere ordinary sense of the word, then we apply the positive pole to the vertebra opposite the highest nerve-origin that can be concerned, and we stroke the negative pole down by the side of the spinous processes, some forty times in succession. The proportion of cases of idiopathic neuralgia in which this treatment succeeds is, according to Benedikt, very large. In other cases, he sends the current from the cord to the apparent seat of pain.

On the other hand, Althaus* tells us that, whether the

---

* "A Treatise on Medical Electricity," second edition, Longmans.

application be central or peripheral, it is the positive pole, alone, which should be applied to the part which we intend to affect: and that the application of the negative pole in this situation is rather likely to do harm than good, as proving too exciting. Eulenburg, also, says that in general the positive pole should be applied to the seat of the disease, the negative on an indifferent spot, or on the peripheral distribution of the nerve.

It is, however, very doubtful to me whether, in the majority of cases, the direction of the current makes any considerable difference in its effects, provided only that the stream is fairly directed so as to include the in the *locus morbi* circuit, and care is taken to apply it with sufficient persistence and with not too great intensity. Upon this point I am glad to be able to cite the authority of Dr. Reynolds, whose experience is very large. This author, while admitting that in theory the "direct" and the "inverse" currents would seem likely to have different effects, declares that in practice this does not prove to be the case, either in the instance of pain of nerve or of spasm of muscle. Dr. Buzzard, also, in relating a very striking case (which I had the advantage of personally observing) before the Clinical Society, particularly mentioned that the direct and the inverse currents had a precisely similar effect in relieving the pain. The patient suffered from severe and probably incurable cervico-brachial neuralgia; the poles were placed, respectively, on the nape of the neck and in the hand of the affected limb, and whether the positive was on the nape and the negative in the hand, or *vice versa*, the effect was the same. Very striking remission of the pain was always produced, and the immunity from suffering sometimes lasted for a considerable time, while no other plan of treatment seemed to have more than the most momentary effect.

My own experience tells the same story very decidedly, for I have on very many occasions obtained great benefit, both by the direct and by the inverse currents, in the same patient. I shall here relate a few instances:

CASE I.—A married woman, aged forty-eight, whose menstrual periods had ceased quietly some six years previously. She was, on the whole, a healthy person, but had suffered from migraine in her youth, and came of a neurotic family. She was attacked with severe cervico-brachial neuralgia, which resisted all treatment for nearly three months, and, on her then trying a month's change of air and absence from medication, became worse than ever. The constant current was applied, from ten (and afterwards fifteen) cells of Weiss's battery, daily for twenty-four days. the pain vanished finally at the end of thirteen days, and the accompanying anæsthesia and partial paralysis disappeared before the treatment was concluded. In this case the negative pole was applied by the side of the three lower cervical vertebræ, and the positive was applied, successively, to three or four different parts of the most intense peripheral pain.

CASE II.—A young lady, aged twenty-four, suffered from neuralgia in the leg. Galvanization (twenty cells Daniell), from the anterior tibial region to the spine was found invariably to cut short the pain. I now reversed the current; the effect was the same. After ten sittings I suspended the treatment, as there had been no attack for three days; but a week later the neuralgia returned in full fury. I resumed galvanization from periphery to spine; after twelve more sittings the attacks had become rare and slight. I continued treatment for eight days longer, during the whole of which time there was no pain. It had not recurred when I saw her fifteen months afterward.

CASE III.—H. G., a footman, aged twenty-three, applied to me at Westminster Hospital, with neuralgia of the first and second divisions of the right trigeminus, of six weeks' standing. The right eye was bloodshot and streaming with tears, the skin of the right side of the nose and right cheek was anæsthetic, the right levator palpebræ was partially paralyzed. Hypodermic injections of morphia proved only very temporarily beneficial. After a fortnight's treatment with this and with flying blisters to the nape of the neck and the mastoid process, I commenced the use of the constant current daily (ten cells, Weiss). The first application (positive on nape, negative on infra-orbital foramen) stopped the pain, and procured fourteen hours' immunity. On the next day I reversed the current; the pain stopped after three minutes' galvanization; it did not recur for four days, during which time, however, I continued the daily use of the direct current. On the sixth day of treatment the patient came to me with a somewhat severe paroxysm, almost limited to the ophthalmic division; it was accompanied by spasmodic twitchings of the eyelid, and copious effusion of altered Meibomian secretion, looking like pus. Galvanization from supra-orbital foramen to nape stopped the pain in five minutes. The next day the patient presented himself, quite free from pain, which had not returned; the conjunctiva was clear, and there was no visible Meibomian secretion. Inverse galvanization was continued for ten days; but no recurrence of the pain took place. The cure was permanent three months later.

On the contrary, we sometimes see complete failure of the current to affect any good whatever; and in these cases the reversal of the current has not, so far, appeared to me to make any particular change in the result. Such was the case with a patient whose history I detailed (along with that of Case I.) to the Clinical Society. She was an ill-fed and over-worked unmarried needle woman, aged thirty; the neuralgia was a most violent double occipital pain, with foci, on each side, where the great occipital nerves become superficial. The current was passed daily, for some days, from one focus to another (necessarily passing through the nerve-roots and the spinal cord), and the positions of the conductors were occasionally reversed; this not succeeding, the current was applied altogether to the spine, the negative pole being placed on the highest cervical vertebræ, but no good effect was produced after a treatment, altogether, of sixteen days.

Notwithstanding these, and a good many similar facts that could be adduced, I should hesitate to go so far as to say that there is never any importance in the direction of the current. In old-standing cases, where there are well-marked *points douloureux* that are exceedingly sensitive, I have found that the application of the positive pole, successively, on the most tender points, the negative being placed on the spine opposite the point of orgin of the nerve, has had a more beneficial effect than any other mode of application.

There are very considerable differences, both as to the best manner of galvanization, and also as to the chances of doing good with it, in the case of neuralgias of different nerves; and, on the whole, I find Eulenburg's conclusions on this matter very just. He indicates sciatica as the affection which is by far the most curable by the constant current; he says that many cases are cured in from three to five sittings, while others require as many weeks, or even months of treatment; and that a total absence of benefit is only seen in rare cases dependent on central causes, or on diseases which are irremovable (like malignant pelvic tumors). On the other hand, he reports that intercostal neuralgia has never been materially benefited by galvanization in his hands. With regard to ordinary trigeminal neuralgias, he speaks strongly of the current as a palliative, but very doubtfully of its power to cure, in genuine and severe cases. In cervico-brachial neuralgia he speaks of it as dividing with hypodermic morphia the whole field of useful treatment, in the majority of cases. In cervico-occipital neuralgia he says it rarely does much good. I shall return to Eulenburg's estimate of its utility in migraine, presently. Let me here say that I am inclined to indorse everything in the above-detailed statements, excepting that I should place a considerably higher estimate on the curative powers of the current in ordinary trigeminal neuralgias. The remedy, like every other, will doubtless fail in a considerable number of those very bad cases which occur in the degenerative period of life; but if anyone desires to see the proof of the power it sometimes exerts, even in extreme cases, he should study the two most remarkable cases treated by Prof. Niemeyer, of Tubingen, and reported by Dr. Wiesner.* The patients

---

* *Op. cit.*

were respectively aged sixty-four and seventy-four, and the duration of the neuralgia had been respectively five and twenty-nine years; in both the pain was of the severest type, and in both the success was most striking. In one of them every possible variety of medication, and several distinct surgical operations for excision of portions of the affected nerve, had been quite vainly tried. The cases are altogether among the most interesting facts in therapeutics that have ever been recorded. Dr. Russell Reynolds has also told me of a case under his own care, in which a lady, who had been the victim, for twenty years, of an extremely severe neuralgia of the ophthalmic division of the fifth, which attacked her daily, and had caused great injury to her general health and nutrition, was not merely benefited, but the affection absolutely removed, at any rate for a long period, by a single application of the current. I have personally seen no such remarkable cases as these; but I have had some extremely severe cases under my care in which the effect of the current was to arrest the pain in a few applications, and procure a remission for several days, or even weeks. And I have had several slighter cases which were as much cured, to all appearance, as any disease can be, by any remedy.

As a general rule, neuralgia of the limbs requires to be treated with a more powerful current than neuralgia of the face (twenty cells instead of ten). In the latter case, indeed, it is necessary to be exceedingly cautious (commencing with five cells), since a current of high power has been known to produce most serious effects upon the deeper-seated organs; the retina has been permanently paralyzed, by too strong a current applied on the face, and still graver dangers attend the incautious use of galvanization of the brain or of the sympathetic, of which we have now to speak.

Galvanization of the brain is a remedy chiefly employed in true migraine, and is certainly very effective in that disease. I have not found it useful to apply the current in the long axis of the cranium, but transmitted from one mastoid process to the other it has proved most useful; and I am glad to find that my experience on this point coincides with that of Eulenburg. But the use of this remedy is highly perilous in careless hands. In working with either Daniell's or Weiss's battery, it is necessary to use at first only three or four cells, and to increase the number only with the greatest caution. The sittings should never last more than half a minute; but the slightest giddiness should make us stop even sooner. On the other hand, the applications ought to be made daily, and usually twice a day. Ten cells (Daniell or Weiss) is the utmost that will ever be required, few patients will bear so much; and, apart from the possibility of more serious mischief, there is nothing which annoys and frightens patients more seriously than the sudden and intense vertigo which over-galvanization of the brain may induce.

Even more ticklish than the galvanization of the cerebral mass is galvanization of the sympathetic. I am not going to raise here the vexed question in physiological electricity as to the possibility of a galvanization the effects of which shall be accurately limited to the sympathetic. The fact is unquestionable, that very powerful and peculiar effects, utterly unprocurable in any other way, can be produced by placing one pole on the superior cervical ganglion (just behind and below the angle of the jaw) and the other on the manubrium sterni. This is a mode of galvanization which has been highly praised, more especially by Remak, and after him by Benedikt, but it has yielded rather disappointing results in neuralgia in my hands. Either I have not observed any distinct effect at all, or, if a current even a very little too strong were applied, I have repeatedly seen most uncomfortable, and sometimes very alarming, symptoms. I shall not easily forget a patient who applied at the Westminster Hospital, suffering from a severe form of facial neuralgia, and who was persuaded to come to my house and have his sympathetic galvanized. I used only twenty cells of Daniell, but the current had not been applied more than a few seconds when the patient fell on the floor, and remained in a state of half swoon for a considerable time. I allude to this and other less dangerous accidents that I have seen follow galvanization of the sympathetic, not with the view to prove that the method is useless in trigeminal neuralgia—I should certainly hesitate to say that, considering the large amount of respectable evidence in its favor—but I think that it is a procedure requiring the utmost caution, and meantime I have not personally found it nearly so useful as the methods already described.

There are sundry special applications of galvanism to particular forms of neuralgia which require a few words of notice. Of electrical treatment in regular angina pectoris I have had no experience; and in the one case of intercostal neuralgia, complicated with quasi-anginal attacks, in which I applied the constant current to the spine and the cardiac region, in the direction of the affected intercostal nerve, no effect was produced. I shall, however, mention the experience of Eulenburg, as he is a sober and dispassionate writer on the effects of electric treatment in general. He says he believes that in the proper use of the constant current we shall discover the chief, possibly the only direct, remedy for angina; and he describes the apparently favorable results he has already obtained in three or four cases. The current was from thirty cells; the positive pole was placed on the sternum (broad electrode), the negative on the lower cervical vertebræ. The alternative method which Eulenburg suggests, but has not, so far, put in practice, is direct galvanization of the sympathetic and vagus in the neck.

The application of the constant current in neuralgic affections of the larynx and pharynx is of most indisputable service; the experience of Tobold* upon this point is fully borne out by my own, as far as it goes. In many cases it will be sufficient to place the positive pole (from fifteen cells Weiss) on the pomum Adami, and the negative on the nape of the neck, and to keep up a continuous current for five or ten minutes daily; but in some cases the direct application of the current to the pharynx or larynx may be required; in such, a modification of Dr. Morell Kackenzie's laryngeal conductor will be found useful. [I shall have occasion, in Part II., to notice the superior action of Faradization in mere hysteric throat-pain, as distinguished from true neuralgia.]

Neuralgia of the testicle can be best treated, if galvanism be thought necessary, by immersing the whole scrotum in a basin of salt and water, in which the positive pole is placed: the negative pole is to be placed on the upper lumbar vertebræ; the current should be from fifteen cells Weiss, and the application should last continuously for ten minutes. In neuralgia of the urethra, I should be inclined to adopt a plan, mentioned to me by Dr. Buzzard, of attaching one conductor to an ordinary silver catheter introduced into the urethra, and placing the other pole upon the perinæum.

Neuralgia of the neck of the bladder I have found to be materially relieved by the constant current from twenty cells passed through from pubis to perinæum; the sittings being rather long. I have also, on one occasion, tried the introduction of a proper *porte-electricite*, insulated, except at the tip; but the result was not superior to that obtained in the other way.

As a general rule, it may be said that electricity, like other local measures which tend to concentrate the patient's attention on the parts, is only to be applied to the genital organs as a last resort. This is, of course, especially true in the neuralgias of these organs in women.

In concluding what will doubtless seem to some English readers an over-long and over-favorable estimate of the employment of galvanism in neuralgias, I must carefully guard myself against the supposition that I consider it a remedy to be applied in all cases, or likely to meet with uniform success, even in the forms of the disease to which it is most appropriate. It is a weapon which I seldom employ in the first instance, for many reasons; the principal of which is the costliness of the proceeding to the patient. Either the physician must personally administer the remedy, daily, often for a considerable

---

* Berlin. klin. Wochensch., 22, 1865.

period, or he must make the patient provide himself with an expensive battery; and in the latter case there is, after all, the unsatisfactory consideration that the application (even after the most careful directions have been given) will perhaps be unskilfully and inefficiently made. On the other hand, it is not desirable to delay the employment of galvanism too long, if other remedies have been fairly tried; and the practitioner will do well to remember the distinctions above laid down as to the varieties of neuralgia in which it is specially likely to prove decidedly and quickly beneficial. More especially in sciatica it would really, with our present knowledge, be a decided neglect of duty were we to allow the disease to run any considerable length without giving the constant current a thorough trial. [I can only briefly refer, here, to the novel mode of galvanization introduced by Dr. Radcliffe, and based upon his ingenious theory, according to which the true effects of the voltaic current upon nerve are the result of the charge of free electricity which it sets up, and not of the current directly. The reader will find the whole argument elaborately worked out in Dr. Radcliffe's recent work on "The Dynamics of Nerve and Muscle," Macmillan & Co., 1871. It will be enough to say, here, that the object to be attained, according to this view, is to replace the neuralgic nerve in its healthy physiological state, by charging it with free positive electricity. The manner in which this is done is as follows: In a case, e. g., of cervico-brachial neuralgia, we place the positive pole as near as may be to the central origin of the affected nerve; the negative pole is held in the hand of the same side, which is immersed in a basin of warm salt and water. In this same basin is another electrode, the wire from which is put in communication with the earth—most conveniently by putting it in contact with a gas-pipe. The patient, and the battery, ought properly to be insulated. The result of this arrangement is, that the free negative electricity is carried off by the earth-wire, and the limb remains charged with free positive electricity. I have had no sufficient experience of this method to give any opinion of its merits, but the inventor thinks it decidedly superior to the ordinary modes of applying the constant current.]

(*f*) The last kind of local remedies for neuralgia of which we have to speak are those by which we seek to mitigate the paroxysm by thoroughly excluding the air from the site of apparent pain. These are chiefly of value in those cases where a distinct inflammation (herpetic or erysipelatoid), or an unusual degree of sensitiveness on pressure, etc., has become developed around the superficial branches of the neuralgic nerve. Very much the best agent of this kind with which I am acquainted is the flexible collodion; in neuralgic herpes and erysipelas the effect of this application, conjoined with the hypodermic injection of morphia (preferably in the immediate neighborhood), is of the greatest possible service in mitigating the pain. In herpes it has this further special advantage, that it prevents the occurrence of sores after the vesicles fall, an accident which otherwise will sometimes happen, and which very much increases the severity and intractability of the consecutive neuralgic pain.

4. Lastly, we have to speak of prophylactic measures, which really ought never to be thought of as a separate matter, but always as an essential and most important part of the treatment of neuralgia. The prophylaxis of neuralgia is divisible into (*a*) measures for preventing the development of the neuralgic habit in those who may be supposed to have a predisposition to it; (*b*) measures between the paroxysms; (*c*) measures to be adopted after the attacks have ceased.

(*a*) The measures that should be taken to avert neuralgia, in those who may be reasonably assumed to be predisposed to it, have scarcely received any consideration at the hands of systematic writers; yet this is a most important subject. The persons in question are children who belong to families known to be infected with tendencies to neurotic diseases, or persons whose daily occupations submit them to peculiarly strong predisposing influences of an external kind. The hostile influences that should be avoided, or at any rate compensated, are of several kinds: (1) Psychical; (2) defects of nutrition; (3) mismanagement of the muscular system; (4) sexual irregularities; (5) over-fatigue of the special senses, and insufficiency of sleep, especially the latter; (6) unhealthy atmosphere and climate.

(1) The psychical influences which must be especially avoided, if we would avert the formation of the neuralgic habit, form a large and somewhat indefinite group, which it is doubtless difficult to deal with satisfactorily. The matter is, however, highly important, and the attempt must be made. And there are, at any rate, some leading principles that I feel justified in laying down with confidence.

We shall best commence the inquiry by directing our attention once more to the fact, so often insisted upon in this work, that the large majority of neuralgic patients carry in them the seeds of their malady from their birth. It has been amply proved that every child born of a family that has shown strong tendencies to insanity, epilepsy, paralysis, etc., etc., ought to be looked on as a neurotic subject, and as a potential sufferer from neuralgia. It has been shown that such children will be exposed, even under favoring external circumstances, to the danger of neuralgia at certain important stages of their physiological history. The earliest of these critical periods is marked by the occurrence of puberty; and it is not till this time that psychical influences, as such, come to have any serious bearing on the formation of the neuralgic habit. Mischief may, indeed, be done to the brain and the general nervous system, by injudicious mental training, at a far earlier period; but this mischief, serious or even fatal as it may be, usually takes some other form than that of neuralgia. It will be necessary, here, to reflect a little upon certain features of the childish mind, in order that we may rightly estimate the kind of influence which puberty exerts upon it.

A very young child is selfish, in the purely animal sense; it is greedily acquisitive, and its selfishness is unchecked by any sense of shame. With later childhood there comes a sense of right and wrong, and a sensitiveness to shame, which check this tendency; still it is the exception rather than the rule to find any great capacity of self-abnegation in young school-boys. But a moderately healthy-minded child, up to the age of puberty, is only acquisitively selfish; he is not self-centered in the sense of dwelling upon his own mental state, and reflecting upon the nature of his motives and feelings. It is with the age of puberty that self-consciousness begins to be a feature in the mind of the young, and its appearance marks the entrance of a dangerous element into the character. It is an inevitable stage in mental growth, and, if wisely dealt with, is ultimately productive, not of evil, but of good; but it is more perilous to some children than to others, and it is especially fraught with danger to those whose nervous centres are, by inheritance, weak and unstable in whole or in parts. The mental antidote to its possible evil effects is to be found in a vigorous (but not excessive) training of the mind in studies which shall be as far as possible external, and the discouragement of all tendencies to introspection. I would venture to express the decided opinion that the common idea, that close study injures the young, is only true in a modified sense. It is, however, unquestionably the fact, that hasty and imperfect cram-work does very seriously impair the stability of the brain and the nervous system in young people; there is a spurious excitement about this kind of learning (especially when it is mainly competitive, and directed to the gaining of prizes and medals) which must be injurious. But I think it is quite ridiculous to suppose that, in this country, the actual amount of intellectual labor undergone by boys and girls at school is sufficient to do harm, were it only regular and systematic, and carried out in a conscientious manner; on the contrary, though I think that the total daily period occupied in study ought not to exceed some six or seven hours, I believe that the insisting on strenuous diligence during school hours, and the maintenance of a high standard as to the quality of the work exacted, is all on the side of nervous health. But, an even more serious and difficult matter than the regulation of the amount of intellectual work to be done is, the question how we are to deal with the unfolding emotional instincts of the boy or girl who has reached the age of puberty. It is useless to ignore this side of the mental life; it will assert itself either for good or for evil. At the risk of seeming to meddle with matters that belong to the school-master rather than to the physician, I would urge very strongly that a portion of the training be deliberately directed to a serious study of one or other of the fine arts—to that one, whether poetry, painting, sculpture, or music, to which the boy or the girl instinctively leans. I am aware that there is a prejudice among parents that the study of the fine arts renders young people idle and indifferent to other branches of education and other duties of life. I believe that this only applies to the miserably inefficient way of teaching these subjects which prevails at present in all but a few English schools; and that, in truth, a thorough knowledge of the principles of either music or painting, and a real study of the best masters, would be sure to prevent the development of that lazy, conceited manner, and that neglect of other duties, which no doubt unfavorably distinguish a good many of the young ladies and gentlemen who dabble a little in music, or painting, or versification. We want the German rather than the English type of training, we want the acquirement of sound knowledge of the principles of music (at any rate) to be made so common that the accidental possession of two pennyworth of superficial accomplishment in that line shall not enable young ladies and gentlemen to give themselves airs in society. The truth is, that the young people who make music or painting an excuse for idleness respecting other matters are invariably imposters. even in that which is their own supposed *forte*. On the other hand, the serious study of art, a certain definite portion of time being set apart for it, and thoroughness being insisted upon, is, I believe, an admirable vent for the emotional effervescence of commencing sexual life; and I no less firmly believe that the things that are usually substituted for it are intensely pernicious. I have already, in the chapter on Pathology, remarked on the mischief which is often done by the anxiety of religious parents to make their children (usually somewhere about this perilous time of puberty) experience the emotional struggle which is believed to end in a change of heart and principles. I need, therefore, only now repeat the expression of my intense conviction that the results of this process, as seen by the physician to occur within that mental region where the emotions and the organic nervous system come into closest relations, are simply disastrous. It is not my business to suggest the proper alternative to a mode of spiritual training which I think deleterious; I can only intimate, in the most general way, my belief that a calm and systematic training in the simplest principles of duty and religion is greatly more suitable to the immature mind and brain of youth than any strong emotional excitement on such topics. But if ill-regulated spiritual emotion of a religious kind be a dangerous thing for young persons in the most serioue crisis of bodily development, far more decidedly pernicious is the spurious excitement of feeling which is directed to lower and often most unworthy objects. The increasing precocity of boys and girls, in their familiarity with the most objectionable aspects of passion and intrigue, is steadily fed, in the present day, by a system that allows them, too often, unlimited access to light literature which (as is strikingly the case with many novels of our day) is at once devoid of true literary and artistic merit. and at the same tIme replete with sensational incident of a vulgarly exciting kind. The same degrading tendency is very distinctly to be noted in the character of the dramatic and other public exhibitions which are most popular at the present day; the main characteristics being, bad art, and thinly-veiled sensuality, all the more pernicious for being veiled at all. It would be a hundred times better that a boy, or even a girl, should study the frank and outspoken descriptions to be found in Shakespeare or Fielding, with all their occasional coarseness, than that they should enervate their minds with the sickly trash that is most current and most popular at the present day. in theatre and circulating library.

(2) The defects of nutrition that assist the development of the neuralgic tendency are often the consequence of a system which, it is to be hoped, is to a large extent becoming effete, but which, nevertheless, survives in sufficicient vigor and extent to demand express reprobation. It was till lately the general, and it is still a too common practice, to keep children and young persons on a very insufficient allowance of the most important elements of food; the state of things in this respect, both in public and private schools, in the first half of the present century, is a lasting reproach to the medical practitioners of those days, who scarcely lifted a finger to amend it, even when they did not expressly approve it, under the influence of absurd theories about the dangers of excessive "grossness of blood." It is indeed amazing that, with the palpable fact staring them in the face, of the rapid and incessant additions to tissues which are being made by children and young people, medical men should have failed to perceive the necessity for supplies of food practically unlimited except by the capacity of digestion. Yet this seems hardly ever to have been thought of, and the unfortunate results seem scarcely to have been noticed, except when they led to emaciation or consumptive disease. But the effects were perhaps even more disastrous where, with a maintenance of a fair amount of muscular nutrition, there was only a little dyspepsia, and perhaps some slight tendency to nervousness, to show that anything was wrong. The children who were born of strong and healthy parents, may have suffered comparatively little from

this regimen as regards their nervous system, but those who were born of neurotic ancestors undoubtedly suffered extensively. The crisis of puberty was, in such ill-nourished children, too frequently the signal for an explosion of epilepsy, chorea, or neuralgia; and too often the mischief was yet further increased by a most injudicious medical treatment, including a deterioration rather than an improvement in the already insufficient dietary system. At the present day, however, we may fairly hope that common sense is prevailing, so as to put an end to this mischief as regards the children of the upper and middle classes. Unfortunately, with the poor a similar ill-nourishment of the young is too often inevitable, and the consequences are constantly to be traced in enfeeblement of the nervous system, of which neuralgia is a pretty common result.

It cannot be two frequently repeated that for those children, more especially those who come of nervous families, any considerable error in this direction has a fatal tendency to awaken the disposition to nervous disease. At every step of the infancy, childhood, and youth of such persons, the most generous allowance of the more nutritive elements of food is of the first importance. At the same time I am entirely opposed to the practice of giving stimulants to any considerable extent, or indeed to any extent, save in exceptional instances. Good meat, bread, milk, butter, fruit, and vegetables, are really the efficacious means of fortifying the nervous system against the impending dangers. With hospital out-patients, for whom we cannot command such diet, our best course, whenever they show signs of deficient nutrition, will be the steady administration of cod-liver oil for a long period.

(3) The true and proper training of the muscular system is among the most important means of antagonizing the tendency to the development of the neuralgic habit. It is a great mistake to suppose that over-training in athletics of any kind is of use; but the sytematic employment of means which tend to make the muscular system hardy and efficient is of very great benefit. The parents of children who may be supposed by inheritance to possess a tendency to neuralgia would do well to study such a methodical series of directions as those which are given by Mr. Maclaren, in his excellent work on physical training. I suspect that the benefit of judicious gymnastics is wrought in two ways: first, by its improving circulation and general nutrition, including the nutrition of the nervous centres; and, secondly, that it gives the nervous centres an education, so to speak, by the variety of difficult co-ordinative movements over which it trains those centres to preside. But unquestionably the matter is a science, not a mere rude art, and requires to be studied as such.

(4) Of unspeakable importance to the object of averting the

formation of the neuralgic habit is the prevention of sexual irregularities in the young. Under this heading is included a large and various group of influences; of these the first that requires notice is the prevention of precocious sexual stimulation, whether by talk or by acts, which may precipitate the occurrence of puberty at an unnaturally early age. I know very well how difficult it is to devise any scheme which really would effectively control and antagonize the worst mischief of schools; but it is at least a duty to say here, that no experienced physician can doubt that such a scheme must be found, if we are ever to hope for a healthier race of children and of young men and women, and if we are to break down one of the most potent of the influences that go to the production and maintenance of the neurotic disposition. I would be clearly understood not to suppose for a moment, either that this sort of cause is usually at work in the production of neuralgia in the young, or that of itself it is sufficient to produce the disease; but I would say, for certain, that on children of nervous families such influences act with disastrous energy; and, moreover, that where we see signs, in a neuralgic young person, of that general form of bad health which is connected with precocious puberty, we may be nearly certain that such influences have actually been at work. At all cost, and by all conceivable means, all children, but most especially the delicate and nervous ones, ought to be shielded from the risk of this occurring.

Another form of sexual irregularity which can be counted as a contributor to the formation of the neuralgic habit is menstrual irregularity, especially at the commencement of sexual life. By far the most mischievous in this way is menorrhagia of the young. I have seen exceedingly severe and intractable neuralgia set up by it. As regards the influence of simple amenorrhœa, I am by no means clear: it seems pretty nearly as likely that the deficient excretion (when not dependent on mechanical cause) is a mere sign of the general weakness which also predisposes to the neuralgia, as that the neuralgia is in any way the direct consequence of the amenorrhœa.

Leucorrhœa, especially when profuse and long-continued, is a much more indisputable factor in many neuralgias. It is a point of real importance to put an end promptly to such a discharge, if it exists, and the usual remedies—cold bathing, mild astringent injections, etc.—should be at once prescribed.

Dysmenorrhœa, a painful menstruation, when not dependent on a purely mechanical cause, affords a strong example of neuralgia connected with sexual difficulty; but there is every reason to think that the neuralgia is the primary and not the secondary affection. The only effective prophylaxis, therefore, is the adoption of such general measures as will raise the whole tone of nervous health. It often happens that marriage completely cures the tendency to these attacks.

(5) Insufficiency and irregularity as to the allowance of sleep are potent influences in developing neuralgia in those who are hereditarily predisposed. It is needless to say a single word to prove the imperative need of the young for periodical and prolonged repose from the conscious actions of the nervous system. Full ten hours of sleep in the twenty-four, for boys and girls who are at or near the period of puberty, is an absolute necessity if we would prevent any existing irritability of the nervous system from developing into the fully-formed neurotic temperament. Indeed, I believe that, for all young people (but especially girls) up to the age of twenty-five, this allowance is not the least beyond what is necessary: only the need is most pressing at, and just before, the devolopment of the sexual organs. Of course a much larger allowance of sleep is necessary in actual infancy: from seven to twelve we may be content if we get nine hours clear sleep; but during the two or three years preceding puberty we should insist upon ten hours, at any rate for children who possess the nervous temperament.

(6) Impurity of the atmosphere in which they habitually or daily reside must be carefully shunned for young children, especially for the nervous. The kind of dull and diffused headache which children often complain of, after study for some time in a close, ill-ventilated school-room, is very likely (if the bad influence be continued for a number of years) to develop itself, at puberity, into a regular migraine. Purity of air in the school-room must therefore be scrupulously provided for; and the same thing must be attended to as regards the sleeping rooms.

Of the climatic influences we may speak in a few words. Besides the avoidance of distinctly malarial districts, and also of places where, although there is no distinct ague, there is a prevalence of neuralgic or even of so-called "rheumatic" complaints, it is necessary very carefully to shun damp soils, and places where there is a great deal of harsh and cold wind. Mere lowness of average temperature is not in itself a strong predisposer to neuralgia, at any rate if guarded against by abundant food and the use of such clothes as will prevent children from ever feeling chilly and depressed. But damp and harsh winds are actively bad; and when joined to habitual or frequent lowness of temperature, they constitute very unfavorable surroundings for the nervous systems of delicate children.

(b) We come now to the prophylaxis which is to be adopted in the intervals of the paroxysms when neuralgia has been actually set up. This consists essentially in three things: (1) Physiological rest, as perfect as possible, of the affected parts; (2) protection from cold; 3 protection from sunlight; 4 avoidance of injurious mental emotions.

1. The maintenance of physiological rest, to the greatest extent that is possible, is an absolute necessity, if we would shield a nerve, which has lately been attacked with neuralgia, from fresh paroxysms. The most evident illustrations of this fact are afforded by those neuralgic affections in which it is most difficult to adopt this precaution. Thus the greatest embarrassment from this cause is met with in the case of sciatica; a mild case is often converted into one of great severity and intractibility because the patient, in the early stages, either cannot or will not maintain the recumbent posture. So, too, though in less marked degree. the cure of cervico-brachial neuralgia is often greatly impeded by the difficulty of maintaining complete rest of the limb. Again, in neuralgia affecting the third division of the fifth, the movements of mastication and of speech are a terrible hinderance to the progress of recovery; and it often becomes necessary, in severe cases, to prescribe absolute silence, and even to feed the patient exclusively with such liquid or semi-liquid food as shall require no efforts of chewing.

2. Preservation from external cold is highly important. When a nerve of the arm, or leg, or trunk, is affected, warm flannel under-clothing ought immadietely to be adopted. The patient who has been suffering from cervico-occipital neuralgia should for some time, in anything but quite summer weather, never go out without wearing a warm comforter round the neck. The sufferer from facial neuralgia should for some time after the cessation of actual attacks never face wind without wearing a thick veil.

3. Exposure to bright light must be scrupulously avoided by sufferers from ophthalmic neuralgia. The affection known as "snow-blindness" is really a neuralgia, with vaso-motor complications, produced by the glare of light reflected from snow; and one of the severest attacks of neuralgia which I personally ever experiened was provoked in this way. Even the comparatively slighter, but for an Englishman unusual, glare of sunlight which one meets with during the first days of a Continental holiday, in wandering about towns made up of clean white stone or whitewashed houses, is enough to provoke an attack, unless the eyes are carefully guarded with colored glasses.

4. It is scarcely necessary, after what has been already said, to insist upon the absolute necessity of mental quietude, as far as this can be obtained. This precaution is more or less important in all neuralgic affections; but in migraine and in other trigeminal neuralgias it is almost of more consequence than any other prophylactic measure; and in angina pectoris it is so essential that adoption or neglect of it may easily turn the scale between life and death. All forms of abdominal visceral neuralgia, also, are greatly affected by emotion, and

passion or strong excitement of any kind must be scrupulously shunned if the neuralgic habit is to be broken through. Unfortunately, it too often happens that the mental surroundings of the patient cannot be so changed as to enable us to carry out this kind of prophylaxis effectually; and neuralgic cases of this class are among the severest trials of the physician's tact and skill, and too frequently defy his efforts.

(*c*) The precautionary measures which are to be adopted, after the neuralgic habit has apparently been fairly broken through, in order to prevent the patient from sliding again into the old vicious groove, can hardly be defined with exactness though their general character will be readily gathered from the picture of the clinical history and pathology of the disease which has been exhibited at large in this work. They mainly consist in the avoidance of severe, and especially of unequal, strains upon bodily or mental powers; and in redoubled carefulness in these respects at those natural crises in the life of the organism which have been shown to exercise so important an influence upon the neuralgic tendency. To a certain extent, also, but with much precaution, we may attempt to modify the pheripheral sensibility by what is commonly called a hardening regimen. Thus, with great care, and proceeding in a very gradual manner, we may by degrees accustom the patient to a larger amount of exposure to free air, and even at last to rough weather, so that in the end he may become less sensitive to some of the commonest immediately exciting causes of neuralgia. If one were to construct an advancing scale of such measures, one might arrange them something like this: First, in-door gymnastics, and gentle horse-exercise for out-door work, in fine weather only; then horse-exercise alternated with pedestrianism, sea-bathing in warm weather; and, finally, we should try to reach a stage at which the patient can well endure a ten or fifteen miles' walk or ride every day, and be comparatively careless about the weather. In reaching this latter stage I have seen some patients helped, in an extraordinary degree, by the frequent use of the Turkish bath, followed by douche. Upon this latter subject I beg to offer some remarks, which are the result of pretty careful and extensive study of the effects of the Turkish bath in a variety of chronic nervous diseases. I believe it to be a very great mistake to suppose that, either in rheumatism or in true neuralgia, the process of the bath should be prolonged to such an extent as is commonly done. Instead of the usual slow heating process, gradually carried to a point at which excessive sweating occurs, I believe that the really scientific is the following: The patient should as quickly as possible get into the hottest atmosphere he intends to expose himself to, which should never be more than about 170° Fahr. He should stay in this place just long enough to get thoroughly hot, and, with the assistance of a glass or so of water drunk, throw himself into a free but gentle perspiration. He should then be rapidly shampooed, exposed to the spinal douche for two or three minutes, and then pass to the cooling-room. Let him beware of too long dawdling in the latter place, and let him avoid smoking there. It is a positively dangerous thing to cool one's self quite down to the normal heat, still more so to induce the slightest chilliness; the body should be still in a universal glow when one issues into the street. Over and over again I have proved upon myself that it is the beneficial method, whereas the prolonged use of the bath, the production of very copious sweating, and above all a lengthened cooling process, most seriously exhaust the nervous energy.

There are certain special considerations as to the habits of life that require a word or two. I need say nothing more to enforce the views already put forward as to the necessity of copious supplies of food. I need only refer to what I have already said about the decidedly mischievous tendency of anything like habitual excess in the use of alcohol, merely adding a special caution against such indulgence during, and particularly toward, the end of the period of sexual activity. There is one more topic upon which something must be said, namely, the extent to which sexual intercourse should be allowed. Speaking of neuralgia generally (excluding neuralgic affections of the sexual organs themselves), it may decidedly be said that the regular and moderate exercise of the function, during the natural period of sexual life, is beneficial; but that excess is always dangerous, and that the continuance of sexual intercourse, after the powers naturally begin to wane, is extremely pernicious in its tendency to revive latent tendencies to neuralgia. As regards neuralgias of the sexual organs, it is very difficult to speak positively; and yet I believe that (once the neuralgic habit broken through by other means) it is very desirable that the patient should live according to the laws of normal physiological life.

NOTE I.

ADDITIONAL FACTS BEARING ON THE QUESTION OF NEUROTIC INHERITANCE.

The following cases must be now added to those recorded in my list of private patients whose family history has been ascertained with reliable accuracy.

CASE I. is that of a gentleman, aged forty-seven, the subject of lumbo-abdominal neuralgia: no history of nervous disease in the family; his mother, however, was of a "nervous" temperament.

CASE II.—A gentleman, aged sixty-four, suffering from

angina. His family nervous history is fearful. On the father's side it is not possible to get a clear account. But on the maternal side there has been a strong tendency to insanity and suicide; and in the patient's own generation one brother committed suicide from insanity, and one sister is still alive, insane. An interesting fact is, that the mother's family have shown an extraordinary proclivity to erysipelas.

CASE III.—The young gentleman, whose single but extremely severe attack of angina is previously described, comes of a family in whom the tendency to neuralgia is undoubtedly very strongly inherited. His father is frequently and very severely *migraineux*, and in early life suffered cardiac symptoms not unlike his son's. A brother was also liable to attacks of true migraine between puberty and the age of twenty-one.

CASE IV.—On the other hand, a case of angina which I saw in the country, last year, occurred in a gentleman, aged fifty, whose family presented no traceable neurotic history. But the damage inflicted upon his nervous system by various external influences was quite extraordinary. In some way or other he got some attacks of migraine at the age of fifteen or sixteen; for these he was treated with bleeding, and with a most savage antiphlogisticism generally. From that time he never got free of the neuralgic tendency. He used to have not only facial, but intercostal neuralgia; for this last he was repeatedly bled, under the idea that it was pleurisy. Added to all this he habitually did an immense deal of brain-work in his study, and for years had performed clerical duties of the most exacting and exhausting character. It is not much wonder that these combined circumstances had sufficed to generate the neurotic temperament.

NOTE II.

THE INHIBITION THEORIES OF HANDFIELD JONES AND JACCOUD.

In the present transitional state of opinion concerning the mode in which the phenomena are produced that are popularly known under the name of "reflex paralysis," I cannot pass without notice the doctrines of these two observers. The reader will have perceived that, as regards the secondary paralytic symptoms observed in neuralgias, I explain the phenomena mainly on the theory of a process which is central, and not perepheral, in origin. And, even where, as in some few instances, it seems possible that the starting-point was an organic affection of some viscus, we must always consider the possibility that the link between this and the neuralgia and paralyses was a neuritis migrans travelling inward to the sensory centre, and from that passing over to motor centres and thus producing paralysis; or that, without the intervention of any truly inflammatory process, the continual impressions streaming in upon the cord from the original seat of organic disease may damage the nutrition of the sensory nerve-root, producing a partial atrophy, and that this process may extend to the motor root.

It remains, however, to inquire whether the influence of powerful peripheral agencies may not, in a purely "functional" manner, disable the nerve-centres for a time, causing paralysis with or without neuralgia. The main supporters of such a doctrine are Dr. Handfield Jones[*] and M. Jaccoud.[†]

Dr. Handfield Jones expressly rejects the theory of Brown-Sequard, as to spasm of the vessels in the nerve-centres, and we need not repeat his arguments on that head, because it seems to be generally felt that the vascular spasm theory will not account for the facts. Jones believes that the state produced in the nerve-centre by the peripheral influence is one of paresis from shock-depression, and that from the sensory centre this state can communicate itself to motor and vaso-motor centres, though commissural fibres. He does not believe in the existence of a special inhibitory portion of the nervous system: he believes that an impression may prove stimulating when it is mild, or paralyzing when it is strong; and that any afferent nerve may convey either the one influence or the other to the centres and thus produce secondary stimulus or secondary paralyses in various efferent nerves. Jones has the distinguished merit of being one of the first authors distinctly to perceive that pain must rank on the same level with paralysis: hence he sees nothing unintelligible in the communication of paralysis to a motor centre from a sensory centre that was in the state which the mind interprets as pain.

The *theorie d'epuisement* of Jaccoud (Erschopfungs-theoric) also denies the possibility of Brown-Sequard's idea of prolongde spasm of the vessels of the centres. It imagines that powerful peripheral excitements exhaust the irritability of the nerve, and through that of the centres, and induce a state of unimpressibility—analogous to that which exists in a nerve or nerve-centre, which is included in the circuit of a constant current. The nervous force is wasted, and, until an opportunity of repose is afforded to the centre, the faculty of impressibility cannot again revive.

I must say that of these two theories I decidedly incline to that of Handfield Jones (though I imagine that in reality the cases are extremely rare, if there be any, in which the change in the centres is really only functional and non-organic), I

---

[*] *Op. cit.*
[†] "Les Paraplegies et l'Ataxie du Mouvement." Par S. Jaccoud. Paris, 1864.

prefer the idea of paralyzing shock to that of exhaustion from over-excitement. from a consideration of the nature of that form of peripheral influence which has been specially mentioned by authors as competent to produce this sort of "reflex" affections, namely, intense and persistent cold. It seems to me a mere abuse of words to speak of this as an agent that could exhaust the nerve by over-stimulation; it must surely exhaust it in a much more direct manner than this, namely by the direct physical agency of withdrawing heat from the nerve, and spoiling its physical texture, *pro tanto*. If such an effect as that which must thus be produced on the nerve, and through it on the centre, is to be looked on as a case of over-stimulated function, then, it seems to me, there is no meaning in language, and no possibility of attaining to clear ideas on the subject of nervous influence.

NOTE III.

ARSENICAL TREATMENT OF VISCERALGIÆ.

Since writing the above chapter on the Treatment of Neuralgia, I have had two fresh and very striking examples, in private practice of the power of arsenic to break the morbid chain of nervous actions in angina pectoris.

The first example was that of a medical man, aged seventy-five, in whom a neuralgia, originally malarial in origin, and of some years' duration, had fixed itself for some time in the fifth and sixth left intercostal spaces, and of late had become complicated with anginoid attacks of an unmistakable character, though not of the highest degree of severity. The case certainly seemed very unpromising, looking at the patient's age and the consequent high probability that there was much arterial degeneration. However, the use of Fowler's solution (five minims three times a day) was commenced and steadily pushed. The anginoid attacks rapidly diminished in frequency and at the end of ten days' time were entirely gone, and after one month of treatment he still had no return of them, although they had previously been of daily occurrence. It is a curious fact, whether a mere coincidence or not I cannot say, that, some few days after the anginoid attacks ceased, he began to experience somewhat severe pains, rheumatic in feeling, but unattended with heat or swelling, in the elbows, wrists, and fingers, symmetrically. This has nearly disappeared, but he is still free from angina. There is no discoverable heart-lesion in this patient.

The other case was that of a fine old man of sixty-four, who, but for some few slight attacks of gout, a few small calculi, and a troublesome prostatic affection, had always enjoyed remarkably good health, until about five months ago, when he began to notice tightness across the chest, etc., when he walked uphill. About a fortnight before he came to me, he was seized with very violent and alarming paroxysms of pain across the the chest and running down both arms, extreme intermittence of pulse, and a sense of impending dissolution. The attack had recurred daily, at the same hour (6 P. M.), ever since; besides which there was an abiding sense of uneasiness in the cardiac region, and a consciousness that the least excitement or exertion would bring on the paroxysm. I put the patient on five minims of Fowler, three times a day, with directions to take ether when the paroxysms came. At the end of the first week there was already much improvement, the paroxysms having been both less frequent and less severe. At the end of a fortnight's treatment he reported that there had been nothing like a paroxysm for the last eight days, although there was still a good deal of uneasiness from time to time. The hour at which the attack was expected passed by absolutely without a trace of angina. It remains to be seen how long this improvement will last, but the altered state of things, and particularly the suddenness of the change, cannot be overlooked, and has very much struck the patient himself. It is now six weeks since he had any paroxysm.

It becomes more and more apparent that arsenic is generally applicable to neuroses of the vagus. In asthma, I have long held it to be the most powerful prophylactic tonic that we possess. It is also an excellent remedy in gastralgia; although I have rather dwelt (in the text of this work) on the action of strychnia in this disease, I would not omit my testimony to arsenic. Dr. Leared has related some exceedingly interesting cases bearing on this point. (See *British Medical Journal*, November 23 and 30, 1867.)

NOTE IV.

INFLUENCE OF GALVANISM ON CUTANEOUS PIGMENT.

Dr. Reynolds pointed out to me the exceedingly curious fact, which I have several times verified, that the constant current, in relieving facial neuralgia, not unfrequently disperses, almost instantaneously, the brown skin-pigment that has collected in the painful region; *e. g.*, near the orbit.

NOTE V.

THE ACTUAL CAUTERY.

A remedy for inveterate neuralgia which of late years I had almost discarded—the actual cautery—has quite recently yielded me very good palliative results in two cases. Its omission from the text of the chapter on Treatment was an accident due to the effect of habit in making one, half unconsciously, reckon this remedy as a " counter-irritant." The longer I practise, however, the more decidedly I am convinced that the actual cautery, if properly applied, does not act as an irritant at all; and this fact was sufficiently in my mind, when writing of irritant remedies, to made me omit the cautery from that section. I should have inserted it under the heading of remedies that interrupt the conductivity of nerves, and thus give the centres temporary rest. The only useful way to apply it is, to make an iron white hot, and very lightly brush the skin over so as to make an eschar not followed by suppuration. The galvano-cautery (Stohrer's Bunsen) is the best for the purpose, but I have made the flat-iron cautery serve very well.

PART II.
DISEASES THAT RESEMBLE NEURALGIA.

CHAPTER I.

MYALGIA.

Of all the diseases which superficially resemble neuralgia, none are so likely to be confounded with it, on a cursory glance, as myalgia. More careful inquiry, however, furnishes, in nearly all cases, ample means for distinguishing between the two affections.

Myalgia is an exceedingly painful affection, and it is also much more common than was formerly supposed. It is to Dr. Inman that we undoubtly owe the demonstration of the frequent occurrence of this malady, and the facility with which it may be mistaken for other, and sometimes much more serious, diseases, with very disastrous results. At the same time, I must express the opinion that this ingenious author has decidedly exaggerated the importance of this local disease at the expense of an unjust depreciation of the frequency and significance of other painful disorders which have their origin within the nervous system.

Myalgia proper includes all those affections which are severally known as "muscular rheumatism" (for the muscles generally), and "lumbago," "pleurodynia," etc. (according to locality). It is essentially pain produced in a muscle obliged to work when its structure is imperfectly nourished or impaired by disease.

The clinical history of the different varieties of myalgia absolutely requires this key for its interpretation; otherwise, the appearance of the sufferers from different kinds of myalgia is so widely dissimilar that we should be exceedingly likely to miss the important features of treatment, which must be applied to them all in common. Nothing, for instance, can be more strikingly unlike than the appearance of the pallid, stunted, under-nourishment cobbler who complains of epigastric myalgia, and that of the ruddy and muscular navvy who suffers from acute lumbago, or the similarly plethoric-looking country commercial traveller, who has been driving in his gig against wind and rain, and complains of violent aching pains in one or both shoulders; yet one and all of these individuals are suffering from precisely the same cause of pain, viz., a temporarily crippled muscle or set of muscles which has been compelled to work against the grain. Why this state of things should invariably be interpreted as sensation in the form of acute pain never absent, but severely aggravated by every movement of the affected part, is a matter beyond our powers of explanation, we must accept it as an ultimate fact for the present.

There is scarcely any need to describe the pain of myalgia, since almost every one has suffered either from lumbago, or from a stiff neck produced by cold. The pain is essentially the same in all cases; it is an aching actually felt either in or toward the tendinous insertions of the affected muscles, and sharply renewed by every attempted contraction of those muscles. The variations in the character and severity of the pains are really entirely due to the greater or the less opportunity for physiological rest which the muscle can obtain. Thus the most obstinate and the most severe, kind of myalgic pain is undoubtedly that of pleurodynia—pain in the intercostal muscles and their fibrous aponeuroses—a fact which depends on the incessant movements which these muscles are compelled to perform in the act of respiration. And next to this in severity and obstinacy are myalgias of the great muscles which are incessantly engaged in maintaining, by their accurately opposed contraction, the erect position of the spinal column and of the head. This, rate of proportional frequency and severity, however must be taken as strictly relative; *i. e.*, it is correct upon the supposition that the different sets of muscles were equally worked and that the state of nutrition was equal in the different parents. It is otherwise when the conditions are reversed. Thus, the unfortunate cobbler or tailor, who sits for long hours in one cramped and bent posture, is continuously exerting his recti abdominales (probably suffering from an under-nutrition common to all his tissues) to a degree perfectly abnormal, and out of all proportion to the functional work he is getting out of any other part of his muscular system. The consequense is, thas he comes to us complaining of acute epigastric, and sometimes pubic, pain, rising to agony when he assumes his ordinary sitting posture, and only reduced to any thing moderate by the most complete extension of the whole trunk in the supine posture.

There is no need to dilate at greater length upon the varieties in the symptoms of myalgia, according as it affects one or another part of the body. We must consider, briefly the different kinds of cause that produce it. The immediate source of the pain being, as we have seen, the sense of embarrassment in a muscle obliged to contract when unfit for the work, we have to ask what are the remoter causes that can produce this special unfitness for the work of contraction. They are three: (*a*) Overlabor pure and simple (*i. e.*, in proportion to the existing bulk and quality of the muscle); (*b*) cold, and especially damp cold, producing a semi-paralyzing effect on the vaso-motor nerves, and causing congestion and sometimes a little effusion among the fibres or within the sheath of the muscle; (*c*) fatty degeneration of muscle which is exposed to inevitable and incessant work. Either of these conditions may so disable the muscle that its unavoidable contractions will set up the myalgic state.

Undoubtedly however there is something further, in the shape of a natural predisposition not yet understood, which makes some patients so much more liable to suffer myalgic pain as a consequence of this sort of influences than other persons are. I am in no condition to decide what the nature of this predisposition is; I feel sure it is heightened by an inherited or acquired gouty tain, but I have seen it in people whom there is no reason to suspect of gouty tendencies. It appears to have no connection with true rheumatism.

Still after all that can be said, myalgia remains a disease chiefly of local origin, and depending for nine-tenths of its causation upon a derangement between the balance of work and nutrition in the muscle.

As regards the diagnosis of myalgia from neuralgia, which is a very important matter, the following are the main points that we should recollect:

Neuralgia.	*Myalgia.*
Follows the distribution of a recognizable nerve or nerves.	Attacks a limited patch or patches that can be identified with the tendon or aponeurosis of a muscle which, on inquiry, will be found to have been hardly worked.
Goes along with an inherited or acquired nervous temperament, which is obvious.	As often as not occurs in persons with no special neurotic tendency.
Is much less aggravated, usually, by movement than myalgia is.	Is inevitably, and very severely, aggravated by every movement of the part.
Is at first accompanied by no local tenderness.	Distinguished from the first, by localized tenderness on pressure as well as on movement.
Points douloureux, when established at a later stage, correspond to the emergence of nerves.	Tender points correspond to tendinous origins and insertions of muscles.
Pain not materially relieved by any change of posture.	Pain usually completely and always considerably relieved by full extension of the painful muscle or muscles.

The treatment of myalgia is not only satisfactory in itself, but often affords, in its results, a very desirable confirmation of diagnosis.

For a very large number of cases, all that is required is (a) to put and keep the affected muscle in a position of full extension, which is only to be changed at somewhat rare intervals; (b) to cover the skin all over and round it with spongio-piline, so as to maintain a perpetual vapor-bath; (c) on the subsidence of the acutest pain and tenderness, to complete the treatment by one or two Turkish baths, to be taken in the manner that I have recommended by speaking of the prophylaxis of neuralgia.

When treatment such as this cures a pain which was greatly aggravated by muscular movement, we may be sure that pain was myalgic and not neuralgic.

The pain, however, is not unfrequently rebellious to such simple remedies as these, more especially when (as in pleurodynia) we are not able to enforce complete physiological rest of the part. When this is the case, we shall find the internal use of twenty and thirty grain doses of muriate of ammonia by far the most effective remedy. In the first very acute stage of a severe case it may be advisable to inject morphia hypodermically; but this is seldom necessary. The muriate-of-ammonia treatment may be usefully accompanied by prolonged gentle frictions, three or four times a day, with a weak chloroform liniment.

When there is visibly a very great deficiency in the general nutrition, we shall often fail to obtain a cure until we have remedied this defect; and accordingly, in the majority of cases of half-starved and overworked needle-women, cobblers, tailors, and the like, who present themselves in the out-patient room, I accompany the above-named treatment with the steady administration of cod-liver oil for three or four weeks or more.

There is one remedy for this pain which I have myself seen used in only a few cases, but which I believe promises exceedingly well for the treatment of obstinate myalgia; viz., acupuncture. I have not even mentioned it as a remedy for neuralgia, for I believe it to be totally useless in true cases of that disease, whether applied in the simple form or in that of galvano-puncture. I think very differently of its use in myalgia; and I venture to believe that it is entirely to cases of this disease that the exceedingly interesting observations of Mr. T. P. Teale, in a recent number of the *Lancet*, apply. Where (after the usual remedies for myalgia have been applied) we are unable to get rid of a deep-seated and fixed muscular pain, I believe it to be excellent practice to plunge two or three long needles deeply into the muscle near its tendinous attachment.

CHAPTER II.

SPINAL IRRITATION.

' I retain this phrase, not because it is an absolutely good one, but because it has become so familiar that it is difficult to dispense with it. We have taken a useful step, however, in separating the true neuralgias from the somewhat indefinite group of diseases to which this title has been given. I think the reader who has carefully studied Part I, of this work will not deny that the latter disorders present a very clear and definite common outline which distinguishes them essentially from the vaguer affections to be described under the present heading.

Spinal irritation, in my sense, includes all those conditions in which, without any special mental affection, and without any single nerve being definitely affected, there are sensations varying between mere cutaneous tenderness, often of a large and irregular surface, and acute pain approaching neuralgia in character, together with fixed tenderness of certain vertebræ on deep pressure. A very large majority of the phenomena are such as would be popularly included (now that they are known not to be of an inflammatory character) under the term "hysterical." That unhappy word crosses our path at every turn in a most embarrassing manner, and yet it can hardly at present be said that we could afford to do without it.

The more typical cases of so-called "hysterical hyperæsthesia" present the following phenomena: Along with the general symptoms of the hysterical temperament (tendency to causeless depression, variable spirits, sensation of globus, semiconvulsive attacks terminated by the discharge of a great quantity of pale, limpid urine) there is commonly a marked superficial tenderness of the surface everywhere, and an exaggeration of reflex irritability. The general tenderness is so far merely cutaneous that deep pressure is ordinarily borne better than the lightest finger-touch. But besides this there are usually one or several spots in which the tenderness is more profound and genuine. There is almost sure to be some point in the spinal column where firm pressure not merely evokes a complaint of pain, but also induces secondary objective phenomena connected with distant organs, such as nausea and vomiting when the cervical vertebræ are tender, severe gastric pain when the dorsal vertebræ are tender, etc. In such cases there is not only spinal tenderness, but very usually also a well-marked tenderness in the epigastrium and the left hypochondrium, the *trepied hysterique* of Briquet. The reader must, however, be warned that the whole of these three tender points may be merely myalgic, and it is necessary very carefully to observe whether local movements do or do not seriously aggravate the pain in them. And, on the other hand, the spinal tender point may be merely the "point apophysaire" of a true neuralgia which exhibits no other symptoms of the so-called hysteric constitution.

The kind of hysteria that is joined with the existence of fixed tender spots in definite points of the vertebral column is not commonly distinguished by the occurrence of cutaneous anæsthesia; but those writers are certainly wrong·in saying that such a combination never takes place. I have seen examples of the most marked union of the two classes of symptoms in the same person.

These cases of so-called spinal irritation with general hysteric manifestations are very commonly attended with paroxysmal pains that approach true neuralgia in character. Nor is it to be denied that we sometimes meet with the combination of general hysteria, spinal tenderness in definite points (with secondary spasmodic or paralytic phenomena always following pressure exerted on the latter), and true neuralgia limited to one nerve. But the more typical spinal irritation cases are merely complicated with a tendency to vague pains which are shifting both in character and position, not with definite unilateral neuralgia always haunting the same nerve and exhibiting more or less of the same type. In fact, as far as one can judge in the absence of any precise information as to the condition of the nervous centers in such cases, it would seem likely that the ordinary cases of spinal irritation differ from the true neuralgias chiefly in this—that the injury, or inherited weakness of organization, or both, which is at the root of the malady, is at once slighter in degree, and spread over a larger tract of the nervous centres, than that which produces a true neuralgia. I believe that Dr. Radcliffe is right in supposing it to be probable that a blow or other injury to the back producing general spinal shock, is the original but unsuspected cause of a large proportion of these cases. One of the most perfect examples of spinal irritation that I have ever seen (and which also contrasts keenly with the commoner hysteric affections on the one hand, and the true neuralgiæ on the other) was that of a girl whom I examined together with Dr. Walshe, Dr. Reynolds, and Dr. Bridge. This young lady was a most intelligent person, and not in the slightest degree inclined to the apathy and idleness so often seen in hysterical people. She had received what was thought at the time to be a very slight contusion in a railway collision, in which, however, her sister, who was in the same carriage, had been severely injured. She nursed this sister assiduously, and it was not till three or four months later that her own health began to fail seriously; but she then became anæmic and extremely depressed. About six months after the accident it was quite casually discovered that there was a spot over the lowest cervical vertebra, pressure on which gave her exquisite pain and a sensation of extreme nausea ; and the very curious observation was made that such pressure instantaneously produced extinction of the right pulse, the left pulse remaining unaltered. In this case it cannot be doubted that a serious shock had been communicated to a lateral segment of the cord involving chiefly the vasomotor nerve fibres, in which probably some decided material lesion had been gradually set up ; and besides this there was probably slighter damage to the spinal cord generally, as there was great general feebleness of movement, though no actual paralysis of the limbs.

Along with the phenomena of fixed spinal tenderness, without distinct neuralgia of any particular nerve, we not unfrequently observe the development of more or less decided tenderness of some of the internal surfaces of the body. I have recently had under my care a young woman in whom a very tender point was developed over the second cervical vertebra, and who suffered from such persistent tenderness of the whole posterior part of the pharynx, that I was for some time seriously apprehensive of the existence of spinal caries and post-pharyngeal abscess. The general character of her symptoms, however, induced me to hope that the case was one of spinal irritation merely, and the event proved that this was the case, for under the use of iron and small doses of strychnia she recovered completely in about three weeks In another patient who came under my care about twelve months ago, there was extraordinary sensitiveness of the gastric mucous membrane, causing exquisite pain after she had eaten almost any thing : there was only occasional vomiting, however, and there had never been any hæmorrhage, so that the evidence for gastric ulcer, which I otherwise inclined to think existed, was insufficient. I discovered that pressure on the third or the fourth dorsal vertebra gave great pain, and produced a strong inclination to vomit ; this made it probable that the affection was spinal, and accordingly all treatment addressed to the stomach was abandoned. Flying blisters to the neighborhood of the painful spinal points quickly relieved all the symptoms.

Another distressing class of symptoms, which is very commonly observed in connection with these cases of spinal irritation, is that of abnormal arterial pulsations : I am not sure whether even severe neuralgia produces more distress than does this pulsation. I have repeatedly seen abnormal pulsation of the carotids in connection with fixed tender-points over the cervical or the upper dorsal vertebræ ; and still more commonly pulsation of the abdominal aorta in connection with tenderness over one or two of the upper dorsal vertebræ. Spasmodic cough and spasmodic dyspnœa frequently accompany tenderness of points in the upper half of the spinal column; and in one instance I have seen pressure on the lowest cervical vertebræ produce a paroxysm which looked alarmingly like angina pectoris. A case of singularly prolonged and obstinate spasmodic hiccough which came under my notice was distinguished by the presence of a fixed tender spot over the third dorsal vertebra.

Prolonged spastic contraction of voluntary muscles, going on, sometimes for weeks, and even months, is a phenomenon that has often been observed; it may attack the arm only, or may affect all the limbs, and the muscles of the trunk and of the neck: it is for the most part symmetrical, but is occasionally unilateral. It begins in the extremities, and is very commonly limited to them; it is much more gentle than tetanic spasm, and is also painless, or nearly so; but the contraction is often strong enough to resist very vigorous efforts at artificial extension.

Paralyses, both of bowel and bladder, have been recorded among the occasional phenomena of spinal irritation with fixed tender points; but I cannot say that I have ever seen such an occurrence. On the whole, I must say that by far the most frequent phenomena of spinal irritation that I have seen have been somewhat diffuse cutaneous or mucous tenderness and irritability (without acute pain) and the presence of tormentating arterial throbbings; also a marked tendency to aggravation of some symptoms, especially the gastric, when firm pressure is made upon the tender spinal points. For a further and fuller account of the phenomena of spinal irritation I may refer the reader to the able article of Dr. Radcliffe,* and the work of the brothers Griffin, already quoted; adding the suggestion, however, that both these authorities, and especially the Griffins, appear to me not to draw a sufficiently clear distinction between the class of cases that I have been attempting to describe and the true neuralgias.

After what has been said, there is no need to draw out a formal list of the points of diagnosis between spinal irritation and neuralgia. It must be admitted, moreover, that the two forms of diseases have a strong connection in the fact that they are each of them most frequently developed in the descendants of neurotic families. It is by the more generalized character of the symptoms, and the absence of the tendency to perpetual recurrence of paroxysmal pain in one definite nerve, that spinal irritation is mainly distinguishable from true neuralgia. I may add that there is a marked distinction, also, in the results of treatment.

The treatment of spinal irritation is, it must be confessed still in an unsatisfactory position; and I believe that a good

---

* Reynolds's "System of Medicine," vol. ii., Art. "Spinal Irritation."

deal of unnecessary discouragement has been occasioned to physicians by their failures to cure supposed neuralgias which really belonged to the spinal irritation class. I would assuredly by no means assert that genuine neuralgia is not frequently intractable, or even incurable; but it is certainly much more curable than spinal irritation; and for this reason, mainly as I believe—that there is much more possibility of aiming our remedies at the actual seat of the disease. On the other hand, in spinal irritation we are confused and distracted with a variety of phenomena for which even the most subtle analysis will frequently fail to trace a common origin. It is true that the existence of definite tender spots in the spine apparently suggests a strictly local application of remedies; and it true also that medication based upon this fact is sometimes very effective; but this is, in my experience, only an occasional result, and the practitioner who trusts to local measures will frequently be disappointed. And, on the other hand, the general tonic treatment, and the use of special medicines, like quinine and arsenic, or the hypodermic injection of morphia or atropia, have nothing like the extensive utility in the treatment of spinal irritation that they possess in that of true neuralgia. Of internal remedies, by far the most useful in my hands have been sesquichloride of iron with small doses of strychnia, and the milder vegetable bitters, especially calumba.

There is one special phase, however, of spinal irritation which is very amenable to the direct, treatment, viz., cutaneous and mucous tenderness. Whatever the "hyperæsthetic" part is within reach, so that we can apply Faradization, we can almost certainly eradicate the morbid sensibility very quickly. The secondary current of an electro-magnetic or volta-electric induction apparatus is to be employed; the conductors should be of dry metal and the negative one, which is to be applied to the painful surface, should be in the form of the wire brush. The positive pole is to be placed on some indifferent spot, and the negative is to be stroked briskly backward and forward over the sensitive skin,. a pretty strong current being employed. The process is painful so much so that it will often be advisable, with delicate patients, either to administer chloroform or to inject morphia subcutaneously before the Faradization. A very few daily sittings of four or five minutes length will generally remove the morbid tenderness completely. Where the tender part is within one of the cavities, at the rectum, bladder, vagina, or pharynx, we must of course use a solid negative conductor of appropriate form, and must content ourselves with applying it steadily to one point after another of the sensitive surface.

The fact that Faridization proves so remarkably useful, in these cases of spinal irritation with diffuse cutaneous or mucous tenderness, is in itself a strong diagnostic between this sort of affection and the true neuralgiæ, which, as I have stated are seldom benefited, and are often made worse, by the interrupted current, though the constant current frequently mitigates or cures them.

Sometimes where it is not possible to apply the remedy directly to the sensitive surface, we may nevertheless do great good by sending the interrupted current through it. Thus, in gastric sensitiveness connected with spinal tenderness in the upper dorsal region, I have seen very great relief afforded by sending a current from the positive pole, placed on the tender vertebræ, to a broad, negative conductor placed on the epigastrium. And similarly, I have seen an acutely sensitive condition of the neck of the bladder greatly soothed by the passage of a current from a painful lumbar vertebra to the perinæum immediately behind the scrotum.

Undoubtedly, however, the more serious cases of spinal irritation will yield only (if they yield at all) to a prolonged treatment in which very skilful use is made of general hygienic measures, and especially of morbal influences. As the brothers Griffin long ago pointed out, although rest is useful in the early stages of this malady, if the disease does not quickly yield to this and to appropriate tonic medication, and perhaps local applications to the spine, it will not do to keep the patient recumbent and confined to the house; on the contrary at whatever cost of immediate discomfort, he (or she for these patients are by far the most frequently females) must be roused up, and persuaded or compelled to take out-door exercise, and if possible to travel, and divert the mind by complete change of scene. When such expensive remedies are out of the question, it seems better that patients, even seemingly very feeble, should take to their ordinary avocations in life again, and fight down the tendency to invalidism. But of course, the decision on such a point must rest with the tact and judgment of the practitioner in each individual case, for there are, doubtless, instances in which the attempt to carry out such a plan, even moderately, would break down the remaining strength, and make matters worse than they were before.

In the worse case of spinal irritation that I ever saw, that of a young lady, aged twenty-eight, there were pronounced anæmia and general feebleness, the true hysteric *trepied* of tender points, painful irritability of the stomach, which baffled all medical advisers and resisted almost every possible form of tonic and nervine medicines, counter-irritation to the spine, and, in fact every thing that one dared attempt with so feeble-looking a patient, but at once cleared up and was quite cured after marriage. And there can be no question that a very large proportion of these cases in single women (who form by far the greater number of subjects of spinal irritation) are due to this conscious or unconscious irritation kept up by an unsatisfied sexual want. In some patients there cannot be a doubt that this condition of things is indefinitely aggravated by the practice of self abuse; but it would be most unjust to think that this is a necessary element in the causation; on the contrary, it is certain that very many young persons (women more especially) are tormented by the irritability of the sexual organs without having the least consciousness of sensual desire, and present the sad spectacle of a *vie manquee* without ever knowing the true source of the misery which incapacitates them for all the active duties of life. It is a singular fact, that in occasional instances one may even see two sisters inheriting the same kind of nervous organization, both tormented with the symptoms of spinal irritation, and both probably suffering from repressed sexual function, but of whom one shall be pure-minded and entirely unconscious of the real source of her troubles, while the other is a victim to conscious and fruitless sexual irritation.

I have already casually alluded to the danger of mistaking mere myalgia for spinal irritation and must again enforce this consideration upon the reader. Myalgic tender points in the region of the spine are common enough; and it would be easy without careful attention, to mistake them for the deeper-seated vertebral tenderness which is truly characteristic of spinal irritation. Hence the utmost care must be taken to ascertain the true history of the commencement of the disorder whether it succeeded to great and long continued fatigue of particular sets of muscles, and whether it is specially aggravated by contractions of those muscles, and relieved by their full extension. The differences of treatment which depend on the diagnosis are too obvious to need dwelling upon.

The question of administering remedies with the direct intention of procuring sleep, for patients suffering from spinal irritation, often becomes an important and a very difficult one. It is, for the most part, highly objectionable to commence the use of such remedies; and yet sleeplessness is a very distressing symptom with many patients, and is, of course in itself exhausting and deleterious. For as long as we possibly can, we should content ourselves with efforts to produce sleep by the timely administration of nourishment. The same general rule of a very generous (though not very stimulating) diet to be enforced as carefully as in the case of sufferers from neuralgia. But it is especially advisable in spinal irritation; that the patient should take some food shortly before bedtime; and it is well, also to place food within reach at the bedside, so that if he wakes up he may take some. If, however, we are absolutely driven to employ hypnoptics, we must commence with the very mildest. The popular remedy of a pillow stuffed with hops will sometimes suffice; and a better way of administering the volatile principle of hops is to scatter a few hops on hot water in an inhaler, and let the patient breathe the steam. Hot foot-baths, with mustard, are also very useful. If these fail, chloral, in moderate doses is probably the best and safest remedy, and, with care not to give too much, we may go on using the same dose without increase for a good many times.

CHAPTER III.

THE PAINS OF HYPOCHONDRIASIS.

There is perhaps nothing, in the whole range of practical medicine, more difficult to seize with clear comprehension, and picture to the mind with accuracy, than the group of pseudo-neuralgiæ which belong to the domain of hypochondriasis. They are among the most indefinable, and at the same time the most intractable, of nervous affections.

To understand what hypochondriac pains are, we must first be familiar with the general character of the hypochondriacal temperament, for the pains are only a subordinate and ever-varying phenomena of the general disease.

Hypochondriasis is not insanity, if by insanity we mean intellectual perversion dependent mainly or entirely on the state of the higher nervous centres. But it is closely allied to insanity in its phenomena, only that these are, as it were, manifested in a scattered form, unequally distributed over the whole central nervous system, and especially affecting the spinal sensory centres. And its radical relationship to true insanity is strongly indicated by the fact that the sufferers from hypochondriasis are nearly, if not quite, always members of families in which distinct insanity has shown itself; indeed, more often than not, of families which have been strongly tainted in this way. In the majority of instances there are psychical peculiarities of a marked kind which accompany or precede the development of the abnormal sensations which form the especial torment of hypochondriacs. Without apparent cause, they begin to evince a heightened self-feeling and an anxious concentration of their thoughts upon the state of one or more of their bodily organs. Or it may be that, before any such definite bias is given to their thoughts, they simply become less sociable and more self-centred, and are subject to fits of indefinite and inexplicable depression, or at least to great variability of spirits. But before long they begin to experience definite morbid sensations, most commonly connected with the digestive organs, and very often accompanied by positive derangement of digestion of an objective character;

such as flatulence, sour eructation, spasmodic stomach-pain, etc. Along with these phenomena, or soon afterward (and not unfrequently before the patient has acquired that intensity of morbid conviction of his having some special disease which is afterward so marked a peculiarity of his mental state), he very often becomes the subject of the kind of pains which it is the special purpose of this chapter to describe.

The pains of hypochondriasis, when they assume any more definite form than that of mere dyspeptic uneasiness, present many analogies with neuralgia. They are not, usually, periodic in any regular manner, but they have the same tendency to complete intermission, and they frequently haunt some one or more definite nerves for a considerable period of time. Of all nerves that are liable to this kind of affections the vagus is undoubtedly the most susceptible; hypochondriac patients very frequently complain of pseudo-anginoid and pseudo-gastralgic pains; next in frequency are nervous pain in the region of the liver, or in the rectum or bladder. The main distinctions by which they are separable from true neuralgia are two: in the first place, the character of the pain nearly always is more of the boring or burning kind than of the acutely darting sort which is most usual in true neuralgia; and, secondly, the influence of mental attention in aggravating the pain is far more pronounced than in the latter malady; indeed, it is often possible, by merely engaging the patient in conversation on other topics, to cause the pain to disappear altogether for the time. But in hypochondriasis it is not often that we are left, for any long time, to these means of diagnosis only; the special character of the disease is that the morbid sensations shift from one place to another, in a manner that is quite unlike that of the true neuralgias. The patient who to-day complains of the most severe gastralgia, or liver-pain, will to-morrow place all his sufferings in the cardiac region, or in the rectum, or will complain of a deep fixed pain within his head; and these changes are often most rapid and frequent. Frequently there are also peculiar skin sensations, which usually approach formication in type, and these, like the pains, are apt to shift with rapidity from one part of the body to another. Later on in the disease, especially in those worst cases which approach most closely to the type of true insanity, there are often hallucinations of a peculiar and characteristic nature, such as the conviction of the patient that he has some animal inside him gnawing his vitals, that he is made of glass and in constant danger of being broken, and a variety of similar absurdities. In short, it is not the full-developed cases of hypochondriasis that need puzzle us, these are usually distinct enough; but the earlier and less characteristic stages in which pain may be nearly the only symptom that is particularly prominent.

In hypochondriasis, as in hysteria, there is often great sensitiveness of the surface; and, as in hysteria, this sensitiveness is found to be very superficial, so that a light touch often hurts more than firm, deep pressure. As in hysteria, too, the tenderness is a phenomenon so greatly affected by the mind, that, if we can divert the patient's attention for a moment, he will let us touch him anywhere, without noticing it at all.

It is a marked peculiarity of hypochondriasis that it is far more common in men than in women; a relation which is precisely the opposite to that which rules in neuralgia. Hypochondriasis is also pre-eminently a disease of adult middle life; it is scarcely ever seen in youth, except as the result of excessive masturbation acting on a temperament hereditarily predisposed to insanity.

The results of treatment frequently assist our diagnosis in difficult cases. Almost any medicine will relieve the pains of the hypochondriac for a time, and it is generally far easier to do him good, temporarily, than it is to relieve a neuralgic patient; but, *en revanche*, every remedy is apt to lose its affect after a little while. The only chance of producing permanent benefit in hypochondriasis is by the judicious combination of remedies that remove symptoms (especially dyspepsia, flatulence, etc.), which mischievously engage the patient's mind, with general tonics, and, above all, which such alterations in the patient's habits of daily life as take him out of himself and compell him to interest himself in the affairs of the world around him. And, after all, our best efforts will frequently lead to nothing but disappointment.

It is notoriously the fact that hypochondriasis especially affects the rich and idle classes; but it would be a great mistake to suppose that it never attacks the poor or the hard-worked: only, in the latter instances, it apparently needs, for its development, the existence of strong family tendencies to neurotic disease, and especially to insanity. Among the numerous debilitated persons who attend the out-patient rooms of our hospitals we every now and then encounter as typical a case of hypochondriasis as could be found even among the rich and gloomy old bachelors who haunt some of our London clubs. I have one such patient under my care now, who has been a repeated visitor at the Westminister Hospital during many years: he has had pseudo-neuralgic pains nearly everywhere at different times; but his most complaint has been of pain in the groin and scrotum of the right side. The existence of what seemed, at first, like the tender points of lumbo-abdominal neuralgia, at one time led me to believe it was a case of that affection; but I was soon undeceived by finding that the tenderness did not remain constant to the same points, but shifted about. This man has professed, by turns, to derive benefit from nearly all the drugs in the Pharmacopœia; but the only remedies that have done him good, for more than a day or two at a time, have been valerian and assafœtida, with the prolonged use of cod-liver oil. He will never be really cured; and I suspect that the secret of his maladies is an inveterate habit of masturbation acting on a nervous system hereditarily predisposed to hypochondriasis.

Sometimes it happens that the starting-point of hypochondriac pains, simulating neuralgia, is a blow, or other bodily injury acting on a predisposed nervous system. Another of my patients at the Westminster Hospital was a policeman, who had received a severe kick in the groin; he suffered pains which at first seemed to wear all the characters of true neuralgia in the pudic nerve, but afterward shifted to other places and exhibited all the intractability of hypochondriasis; the patient also developed the regular appearance and the characteristic hallucinations of the latter disease. On the last occasion when I saw him, he struck me as likely to become really insane, in the melancholic form ; and the probability is that the casualty which he suffered was only accidentally the starting-point of a malady which was inherent in him since birth, and would have been developed, in any case, at some period of his life.

CHAPTER IV.

THE PAINS OF LOCOMOTOR ATAXY.

Considering the vast amount that has been written about this disease during the last few years, it might be thought superfluous for me to give any description of its general features. But it unfortunately happens that there is still great divergence of opinion among authorities as to the true limitation of the group of cases that can properly be ranked under this title, and, indeed, as to the propriety of employing the title at all. The phrase ataxie locomotrice progressive, as every one knows, was applied by Duchenne de Boulogne to a class of cases which really only form a subdivision of the group known under the older title of *tabes dorsalis* and the most advanced German pathologists maintain that the old word was better, and that Duchenne was altogether wrong in making the one symptom, ataxy of locomotion, the bases of a new phraseology;* more especially as his theory as to the seat of the morbid changes was undoubtedly erroneous.

---

* The most complete and careful work of the German school, on this subject, is the "Lehre von der Tabes dorsualis," of E. Cyon. (Berlin, 1867.)

In this country, however, there is as yet no disposition to give up the phrase locomotor ataxy, and it only remains to define with sufficient care the class of cases to which the word is here meant to apply. The disease is understood to depend upon a degeneration of the spinal cord, of which the following description is given by Lockhart Clarke :* "In true locomotor ataxy, the spinal cord is invariably altered in structure. Its membranes, however, are sometimes apparently unaffected, or affected only in a slight degree ; but generally they are much congested, and I have seen them thickened posteriorly by exudations, and adherent, not only to each other, but to the posterior surface of the cord. The posterior columns, including the posterior nerve-roots, are the parts of the cord which are chiefly altered in structure. This alteration is peculiar, and consists of atrophy and degeneration of the nerve fibres to a greater or less extent, with hypertrophy of the connective tissue, which give to the columns a grayish and more transparent aspect ; in this tissue are embedded a multitude of corpora amylacea. Many of the blood vessels that travel the columns are loaded or surrounded to a variable depth by oil-globules of various sizes. For the production of ataxy, it seems to be necessary that the changes extend along a certain length. from one to two inches of the cord. The posterior nerve-roots, both within and without the cord, are frequently affected by the same kind of degeneration, which sometimes extends to the surface even of the lateral columns, and occasionally along the edges of the anterior. Not unfrequently the extremity of the posterior cornua, and even deeper parts of the gray substance, are more or less damaged by areas of disintegration. The morbid process appears to travel from centre to periphery, that is, from the spinal cord to the posterior roots. In the cerebral nerves, on the contrary, the morbid change seems to travel in the opposite direction, that is, from the periphery toward the centres. From the optic nerves it has been found to extend as far as the corpora geniculata, but seldom as far as the corpora quadrigemina. With the exception of the fifth, seventh, and eighth pair, all the cerebral nerves have occasionally been found more or less altered in structure."

The symptoms which occur in cases in which the above are the morbid appearances found after death are (roughly speaking) as follows :† "A peculiar gait, arising from want of co-ordinating power in the lower extremities, a gait precipitate and staggering, the legs starting hither and thither in a very disorderly manner, and the heels coming down with a stamp at each step."

---

* *Lancet*, June 10, 1865. (Comment on a case of Dr. J. Hughlings Jackson's.)
† Radcliffe, in "Reynold's System of Medicine," vol. ii.

No true paralysis in the lower extremities or elsewhere. Characteristic neuralgic pains, erratic paroxysmal in the feet and legs chiefly—pains of a boring, throbbing, shooting character, like those caused by a sharp electric shock.

More or less numbness, in the feet and legs chiefly, in all forms of sensibility, excepting that by which differences of temperature are recognized.

Frequent impairment of sight or hearing, one or both.

Frequent transitory or permanent strabismus or ptosis, one or both.

No very obvious paralysis of the bladder or lower bowel.

No necessary impairment of sexual power.

No tingling or kindred phenomenon.

No marked tremulous, convulsive, or spasmodic phenomena.

No marked impairment of muscular nutrition and irritability.

No impairment of the mental faculties.

Occasional injection of the conjunctivæ, with contraction of the pupils.

The probable limitation of the distinctive phenomenon of locomotor ataxy (the want of co-ordinating motor power) to the lower extremities.

The above description includes all the necessary facts for the recognition of the disease, except one, namely, that the use of the eyesight is always needed in order to prevent the patient from falling during progression ; and is usually necessary even to enable him to stand upright without falling.

The pains of locomotor ataxy are early phenomena in most cases, and they are usually present, more or less, throughout the course of the disease.

They are often preceded by strabismus, with or without ptosis ; the strabismus, is usually accompanied by amblyopia. It may happen, however, that neuralgic pains are, for a considerable time, the only noticeable phenomena ; or they may be attended with a certain amount of anæsthesia.

The most frequent type of the pains is lancinating or stabbing ; they are like violent neuralgias occurring successively in various nerves ; shifting about from one to another. Sometimes it will happen that the pain remains fixed to one particular nerve for hours together ; but it never continues long without showing the characteristic tendency to move about. Most commonly our diagnosis is soon assisted by the occurrence of a greater or less degree of ataxy. But, even before the setting in of definite atactic symptoms, the shifting character of the pains, and the development of a very noticeable amount of anæsthesia, together with the absence of anything like positive motor paralysis, will have given us the necessary clew.

The effect of treatment, or rather its want of effect, usually affords powerful assistance in distinguishing the pains of locomotor ataxy from those of true neuralgia. Even where the pain has been fixed for some hours in a single nerve, and has been stopped by some powerful remedy (such as hypodermic morphia), it will be apt speedily to recur, and frequently in some quite distant nerve.

Locomotor ataxy is a disease affecting chiefly the male sex, and occurring in the immense majority of cases between the thirty-fifth and the fiftieth year.

Not merely is it strictly limited to individuals who belong to families with neurotic tendencies, but it is itself frequently seen to occur in several members of the same family, and sometimes of the same generation. When, therefore, we meet with neuralgic pains of the shifting type above described, it is very important at once to make careful inquiries whether any members of the family have suffered from symptoms of ataxy going on to a fatal result. Otherwise, we might be the more readily deceived into the idea that the pains were merely neuralgic, because the symptoms of the disease are not unfrequently provoked by such causes as fatigue and exposure to cold or wet, which are also very ordinary exciting causes of true neuralgia.

CHAPTER V.

THE PAINS OF CEREBRAL ABSCESS.

Cerebral Abscesses is, fortunately, a rare disease; but the very fact of its rarity makes the resemblance of the pain it causes to that of neuralgia the more likely to lead us into serious errors. We are apt to forget the possibility of suppuration of the brain on account of its infrequence.

Pain in the head is present as an early symptom of abscess in the brain in a large proportion of cases in which there is pain at all. [Of seventy-five cases of cerebral abscess analyzed by Gull and Sutton (Reynolds's "System of Medicine," vol. ii.), pain was a symptom in thirty-nine, and most frequently an early symptom.] Many cases are recorded in which it preceded every other morbid sign by a considerable period. It is usually more or less paroxysmal, often strikingly so; in the latter case, it bears a great similarity to neuralgia. On the other hand, it sometimes takes the shape of a fixed burning sensation, much less resembling neuralgia. The situation of the pain by no means always, nor even usually, corresponds to the situation of the cerebral abscess; on the contrary, abscess in the cerebellum has often caused pain referred to the anterior part of the head, and so on. So long as the disease remains characterized only by pain, more or less, of a paroxysmal character, the diagnosis must be very uncertain; but in the great majority of cases certain more distinctive symptoms soon become superadded; either convulsions (sometimes hemiplegic), vertigo, coma, paralysis, vomiting, or a combination of some of these.

In the stage in which there is as yet no conspicuous symptom but severe pain, the diagnosis of cerebral abscess from neuralgia must rest on the following points of contrast:

Cerebral Abscess.	Neuralgia of Head.
Often occurs secondarily to caries of internal ear, and purulent discharge the result of scarlet fever, measles, etc., in childhood.	Rarely appears before puberty.
Frequently follows a blow or injury.	Comparatively seldom caused by blow, or other external injury or caries of bone.
No true "points douloureux."	If severe, soon presents, in most cases, the "points douloureux."
Usually the pain does not completely intermit.	Intermissions of pain complete, and of considerable length.
Pain often excruciating from a very early period.	Pain usually not very violent at first.
Pain often limited in situation, seems deep-seated, though, as often as not, it has no relation to the site of the abscess.	Pain superficial; follows distribution of recognizable nerve-branches belonging to the trigeminus or the great occipital.
No well localized vaso-motor or secretory complications.	Usually there are lachrymation, congestion of conjunctiva, or other vaso-motor and secretory complications, such as are described in Chapter III.
Very rare in old age; then usually traumatic.	Severe and intractable neuralgia is commonest in the degenerative period of life.
Relief from stimulant narcotics very transitory.	Relief from opium, etc., is much more considerable and permanent.

The only case of cerebral abscess that I have personally seen, in which the above points of distinction would have been insufficient, was that of a boy of sixteen, in whom the only discoverable symptom, for nearly three months, was pain, very strongly resembling ordinary migraine, recurring not oftener than once in ten days or a fortnight, lasting for some hours at a time, and nearly always ending in vomiting, and disappearing after sleep. At the end of the three months, acute pain in the left ear set in, and this was followed, soon, by right hemiplega, coma, and death. It was then discovered, although it had formerly been denied, that the boy had suffered from discharge from the left ear, following a febrile attack which had been marked by sore-throat, and followed by desquamation of the cuticle—evidently scarlet fever. In all cases of severe pain in the head, it is a golden rule to inquire most carefully as to the possible existence, present or past, of discharge from the ear, or other signs of caries of the temporal bone; and, even if no positive history of this kind be given, we should still regard with great suspicion any case in which there has been scarlet fever followed by deafness.

CHAPTER VI.

PAINS OF ALCOHOLISM.

A very important class of pains, which are occasionally confounded with true neuralgias, are those which occur in certain forms of chronic alcoholism. The diagnosis of their true nature is a matter of the utmost consequence, and the failure to recognize them for what they are may have very disastrous results. It is a curious fact that this consequence of chronic alcoholic poisoning has been entirely overlooked by some of the best known writers on that affection; it has, however, been described by Mr. John Higginbottom, and also by M. Leudet.

It must be clearly understood that the pains of which we are now to speak are not among the common consequences of chronic excess in drink. The affections of sensation which most usually occur in alcoholism take the shape either of anæsthesia, or of this combined with anomalous feelings partaking more or less of the character of formication. Chronic drinking has also a tendency, in its later stages, when the nutrition of the nervous centres has been considerably impaired by the habit, to set up true neuralgia, of a formidable type, in subjects who are hereditarily predisposed to neuroses. But the affection of which I now speak may occur at any stage except the very earliest, and, though often severely painful, is essentially different both in its seat and in its general characters, from neuralgia proper.

The earliest symptoms from which the patient usually suffers in these cases are insomnia, and intense depression of spirits, which, however, is not incompatible, indeed is frequently combined, with a morbid activity and restlessness of thought. There is generally marked loss of appetite, but often there is none of the morning nausea so characteristic of the common forms of alcoholicism. Nor is there, ordinarily, any special unsteadiness of the muscular system. The pains are usually first felt in the shoulder and down the spine ; but as the case progresses they especially attack the wrist and ankles; and it is in these latter situations that I have found them to be most decidedly complained of. Their similarity to neuralgia consists (*a*) in their somewhat proxysmal character; (*b*) in their frequently recurring at about the same hour of the day, most commonly toward night; and (*c*) in their special aggravation by bodily and mental fatigue.

Their differences from neuralgia are —(*a*) that they never follow the course of a recognizable single nerve; (*b*) that they are nearly always present in more than one limb, and usually in both halves of the body, at the same time; and (*c*) especially, that they are far less promptly and effectually relieved by hypodermic morphia than are the true neuralgias; indeed, opiates very frequently only slightly alleviate the pain, while they excite and agitate the patient and render sleep impossible. On the contrary, a large dose of wine or brandy will never fail to procure temporary comfort and induce sleep, at least until the patient reaches an advanced stage of the disorder, and is, in fact, on the verge of delirium tremens.

I am not quite sure that I am right in believing that there is a special physiognomy for this form of chronic alcoholism, and yet I am much inclined to believe that there is. All the patients whom I have seen suffering with it have presented a peculiar brown sallowness of face, and a general harsh dryness of the skin, which has usually lost its natural clearness. not only in the face, but even more remarkably in the hands, which are so dark-colored as to appear as if they were dirty. There is usually considerable leanness of the limbs, and, though the abdomen may be somewhat prominent, this does not seem to depend much on the presence of fat, but rather on relaxation of the abdominal muscles, and sometimes flatulent distention of the stomach and intestines. The hands are usually hot, sometimes quite startlingly so.

Some of the patients suffer, besides the pains in the limbs (which they often describe as resembling the feeling of a tight band pressing severely around the ankles or wrists), from frequent or occasional attacks of genuine hemicrania; such a combination is to me always a suspicious sign, and induces me immediately to direct my attention to the possibility of chronic alcoholic poisoning. Otherwise, the limb-pains are often spoken of as resembling rheumatism, but there is no swelling of joints, and usually no decided tenderness of the painful parts. The patient has usually a particular worn and haggard appearence, complains of intense fatigue after the most moderate muscular exertion, and is usually utterly indisposed to physical exercise even though the mind, as already said, may display a feverish activity.

So far as I have seen, the subjects of this affection are by far the most frequently women; and I am inclined to attribute this predisposition of the sex not to inherent peculiarities of female organization, but to the fact that a much larger proportion of intemperate women than of intemperate men indulge in secret excess. They never get drunk, probably, but they fly to the relief of alcohol upon every trivial occasion of bodily or mental distress; and this habit may have been going on for years before it comes to be suspected by their friends or their medical attendant. Meantime, they have been more or less looked upon, and have looked upon themselves as, "debilitated" and "neuralgic" subjects, and have come, either with or without mistaken medical advice, to consider free stimulation as the proper treatment for the very ailments which have been produced by their own unfortunate habits. I cannot avoid the expression of the misgiving, that imperfect diagnosis, and consequent erronous prescription, have done great harm in many such cases. It has happened to me no less than three times within the last six months to be called to lady patients, all suffering from alcoholism induced by a habit of taking stimulants for the relief of so-called neuralgic pain; and in the most distressing of these the mischief had been greatly aggravated by a perscription of brandy, based on the erronous idea that the pains were truly neuralgic. I have already protested against this kind of medication, even in cases that are truly neuralgic in character; but it is doubly mischievous where given for a state of things which actually depends on alcoholic excess.

It is undoubtedly very difficult, sometimes, to elicit the truth, even in cases where we may entertain considerable suspicion that alcoholic excesses are the real cause of the pains which the patient calls neuralgic ; more especially where the patient is aware that he or she is taking an amount of alcohol which is seriously damaging to health. And it is therefore necessary to look out for every possible additional help to our diagnosis. Besides the cardinal features of the disease—the insomnia, loss of appetite, foul breath, haggard countenance, and pains encircling the limbs near the joints rather than running longitudinally down the extremities—there are certain moral characteristics of the patient that often tells a significant tale. The drinker, especially if a woman, is shifty, voluble, and full of plausible theories to account for this and the other phenomenon. It will be well to try the effects of a somewhat sudden though not uncourteous remark, to the effect that the diet should be strictly unstimulating. If this be introduced with some abruptness, in the course of a conversation not apparently leading to it, the patient's manner will not unfrequently betray the truth; while, if our suspicions are groundless, we shall also probably perceive that, in the unconscious, or frankly surprised, expression of the countenance. We may sometimes derive crowning proof of the existence of alcoholic excess by cautious questions which at least reveal the fact that the patient suffers from spectral hallucinations; this is a far commoner occurrence in chronic alcoholism than is generally supposed; it needs to be inquired for with great tact, but, when established beyond doubt, and joined to insomnia and the peculiar foul breath, is of itself sufficient to establish a positive diagnosis of alcoholic poisoning.

The results of treatment, in true neuralgia and in alcoholic pains, respectively, establish an important difference between these affections. In the former malady, for instance, the hypodermic injection of morphia always produces striking palliative, and very often curative effects. In alcoholic pains this remedy either affords only trifling relief, or more commonly aggravates the malady by increasing the general nervous excitement; and the only true treatment is at once to suspend all use of stimulants, to administer quinine, and to insist upon a copious nutrition. If any hypnotic must be employed, let it be chloral, or bromide of potassium with cannabis Indica. It will be well also to put the patient upon a somewhat lengthened course of cod-liver oil. There is one special symptom from which the chronic alcoholist often suffers acutely, namely a hypersensitiveness to cold; for this I found the use of Turkish bath two or three times a week, for three or four weeks, very useful in one case that was under my care. It will be important to insist that the patient shall take the bath only after that shorter method which I have described in speaking of the prophylaxis of true neuralgia.

CHAPTER VII.

THE PAINS OF SYPHILIS.

Syphilis, as has already been shown in Part I. of this work, may excite true neuralgia in subjects already predisposed to the latter. The case of Matilda W., previously given, is an example. The pains, however, which are now to be described, are those which occur in the ordinary course of a constitutional syphilitic infection, and have nothing to do with neuralgia proper, from which they should be carefully distinguished.

There are two varieties of syphilitic pains proper, which are quite distinct. The first kind is represented by the so-called *dolores osteocopi*, which occur in the early stages of the constitutional affection, coincidently with, or just before, the secondary skin-eruptions. The second kind are those which occur in the tertiary stage, and are the immediate precursors of the formation of periosteal nodes.

It is the first of these varieties of syphilitic pains which is least commonly confounded with neuralgia. The pain is referred to the superficial bones, of which those most frequently attacked are the forehead, sternum, clavicle, ulna, and tibia, pretty much those selected for the growth of nodes at a later stage of the disease. Besides the bones, the shoulders, elbows, and nape of the neck are attacked sometimes simultaneously, sometimes successively. The pains are readily controlled by proper treatment; if untreated, their course is very uncertain. When they manifest themselves at the outset of the disease, they usually cease when the cutaneous eruption is fairly out. Commonly, there is no swelling or heat at the painful places; but, when the pains are very severe, nodes now and then form at this early period.*

These early syphilitic pains, in their violent aching character, and their intermittence, occasionally resemble true neuralgia very closely; but they are usually distinguished from it by their symmetrical disposition and by their attacking several bones at once. Moreover, they nearly always show the peculiarity of being distinctly aggravated by the warmth and repose of bed even if they be not altogether absent (as is not unfrequently the case) when the patient is up and moving about. A typical case of this kind is not so likely to be confounded with neuralgia as with rheumatism; but we occasionally meet with cases in which the pains are localized in a manner much more resembling the former. Thus I have met with several instances in which a patient, entirely unconscious (or professing to be unconscious) of having been syphilized, complained of violent pain in one tibia, recurring every night at a certain hour, and at first undistinguishable from that variety of sciatica in which the pain is principally felt in this situation, especially as it was relieved by firm pressure, just as neuralgia is in the early stages. And in one remarkable case, which came under my care at Westminster Hospital, the resemblance to clavus was most misleading:

H. A., aged nineteen, worker in a laundry, presented herself on account of a violent pain in the right parietal region, recurring three times daily with great regularity. The first two attacks occurred in the day-time, the third, which was always the severest, woke her out of sleep about midnight; the pain

---

*Berkeley Hill, "Syphilis and Local Contagious Disorders," p 153.

of this last was so agonizing that on more than one occasion she had become delirious. The girl (whose respectable appearance was against the notion of syphilis) was very anæmic; not, however, with the tint either of anæmic from hæmorrhage, or with that of chlorosis, exactly. It was rather a dirty sallowness of skin; but the gums and the conjunctivæ were exceedingly bloodless, and she complained of almost constant noises in the head. Menses scanty but regular. There was a soft anæmic bruit with the first sound at the base of the heart. Having failed to make any impression on the pains with iron and with muriate of ammonia in large doses, I was led to observe the fact that there was no diffuse soreness of the scalp, such as very commonly occurs in clavus, in the intervals of the pains, and the mere fact that there was this unusual circumstance in the case led me to reconsider the diagnosis thoroughly. In order to be sure of not omitting a point, I inquired, though without any expectation of an affirmative answer, as to the possibility of syphilitic disease; the girl at once confessed to having had sores, and examination detected a papular rash about the shoulders and back and on both thighs. Small doses of mercury greatly relieved the pain within a week, and cured it in less than three weeks; and it was very remarkable that the anæmia, which had obstinately refused to yield to iron, improved at once as the mercury began to relieve the pains. The eruption disappeared simultaneously.

It is the later pains of syphilis, however, that are most frequently confounded with neuralgia, and occasionally with very disastrous results. These pains, which are the precursors of the formation of true nodes, frequent the same localities as those affected by the earlier pains; they may exist in considerable severity for days, or even for many weeks, before any node-formation can be detected. The situation in which, of all others, they are likely to be mistaken for neuralgia is the scalp or face, especially when a single spot is affected on one side, and in the situation of one of the usual foci of trigeminal or occipital neuralgia. I have personally known the mistake to be made with syphilitic affections causing pain, respectively, in the superciliary region, in the malar bone, the jaw near the mental foramen, and the parietal eminence.

The possibility of mistaking tertiary syphilitic pain for neuralgia is fraught with such grave dangers, that we ought to be constantly and most vigilantly on the watch against it. But most especially is this the case when the pain is situated in some part of the cranium, as the parietal or temporal eminences, the mastiod process, or the prominences of the occipital bone. For it must be remembered that the same process, which forms syphilitic nodes upon the external surface of bones, or within bony canals, can produce them on the lining membrane of the skull, with most serious consequences, should the symptoms be neglected or misunderstood.

The pains produced by nodes upon the internal surface of the cranium are usually of a very intense character, and are mostly continuous, though aggravated from time to time, especially at night. Where syphilitic inflammation is diffused over a considerable portion of the meninges, it is certain very quickly to produce symptoms which can hardly fail to apprise us of the gravity of the affection; there will be decided and rapidly increasing impairment of memory, and general cloudiness of intellect, tending toward complete imbecility, the special senses will be greatly interfered with or lost, and muscucular paralysis will be developed. But in the case of a more limited syphilitic affection of the dura mater, pain, of the kind already described, may be for some days the only very noticeable symptom. The following is an instance:

J. E., aged forty-seven, a street and tavern singer, applied to me (November 14, 1861), on account of severe pain in the right temporal region, which had on the whole the character of neuralgia, though rather more continuous than such pain usually is. He said that it commenced on the 10th, without any particular provocation that he knew of, and that it had hardly left him at all from that moment. It kept him awake at night, and that circumstance seemed to account sufficiently for a very worn and depressed look which he presented; he was otherwise a robust-looking man, and at first denied having suffered from any previous illness. The pain always came to a climax about one o'clock, A. M., waking him out of his first sleep in agony, and allowing him little rest for the remainder of the night; toward morning he would drop to sleep for an hour or so. There was no particular tender point, corresponding to any recognized neuralgic focus, yet the pain was limited most strictly to a spot that might be covered with two finger-points. There was no lachrymation nor conjunctival congestion, and nothing to remark in any way about either eye. The patient was ordered quinine in large doses, in the belief that the pain was neuralgic. On the following day he reported himself a trifle better, though still suffering greatly; and on the afternoon of that day there was an almost complete intermission of the pain for several hours; but it returned severely at the usual nocturnal period. On the 16th, at 10 A. M., he came to my house looking exceedingly ill, but the only additional symptom that I could detect was a small droop of the right eyelid. He was subscutaneously injected with one-fourth of a grain of morphia and sent home, where he immediately fell into a heavy sleep that lasted till bedtime. He awoke, undressed himself without feeling much pain, and got to bed; after an hour or so of dozing he was awakened by the pain, which was exceedingly severe. On the 17th he called on me in the morning, and I at once perceived that the ptosis of the right eyelid was much greater, and the right pupil was much dilated and insensitive, and the external rectus was paralyzed; the man also wore a look of stupidity, and answered questions with an apparent mental effort. I now cross-questioned him more closely; and also explored the tibiæ and other superficial bones: on the sternum a distinct though not very advanced node was found. Upon this he was induced to confess that he had suffered from chancre three years and a half previously, and subsequently had "blotches" on the skin, which had quickly disappeared under treatment, of which all that could be learned was, that it was fluid medicine and did not make his mouth sore. He was immediately ordered to take two grains of calomel in pill, with a little opium, every four hours. He had only taken one dose when I was sent for to him, and found him in an epileptiform convlusion, in which the left side of the body was almost exclusively affected; the convulsions recurred several times during the next twenty-four hours, and in the intervals he remained almost completely unconscious. The mercurial treatment was pushed, in the form of calomel-powders placed on the tongue. On the evening of the 18th he began to recover consciousness, and then had a little natural sleep; the next morning, at 10 A. M., he was found to be fully conscious, had had no return of convulsions, but the left arm and leg, especially the latter, were almost entirely powerless; the parietal headache had vanished; the gums were slightly tender; the third and sixth nerves of right side were completely paralyzed. Mercurial treatment was very gently continued, so as to keep the patient on the borders of pytalism for the next three or four days; and he was then put on full doses of iodide of potassium. The pain never recurred; the left extremities recovered power rapidly; but it was six weeks before the ocular paralyses were completely well.

Late in the autumn of 1865 I was sent for hastily one evening to see this same man, and found him totally unconscious and apparently again hemiplegic, but now on the right side. He was miserably wasted, and covered with a rupious eruption; I was informed that he had been leading a most debauched and drunken life for some time past, and that, after looking extremely ill, and apparently half imbecile for a week or two past, he had suddenly fallen down unconscious in the street a few hours before I saw him. He remained deeply comatose, and died the next morning; no *post mortem* could be obtained.

The true neuralgias in which syphilis only plays the part of secondary factor, and which have been referred to in Part I. of this work, may depend for their exciting cause on local syphilitic processes, affecting either the peripheral distribution, the main trunk or the central origin of a sensory nerve; but I have pointed out the fact that, whatever the reason may be, syphilis does but rarely attack the central portions of individual sensory nerves, in comparison with the frequency with which it attacks individual motor (cranial) nerves. But without any neuralgic predisposition at all, and without any limitation of the syphilitic process to a particular sensory nerve, the latter may become neuralgic in consequence of being involved in extensive intra-cranial or intra-spinal syphilitic mischief. The trigeminus is liable to suffer in this way from spreading syphilitic processes about the base of the brain; and my own impression is, that the cause of the neuralgic pain in some such cases is the extension of the mischief to the vertebral artery of the affected side, leading to interfering with the nutrition of the trigeminal nucleus in the medulla. A very interesting case is reported by Dr. Hughlings Jackson (who has done so much to acquaint us with syphilitic affections of cerebral arteries) in vol. iv. of the "London Hospital Reports," pp. 318-321. The patient was a woman, aged twenty-seven, and the initial symptoms of the malady which destroyed her life were violent trigeminal neuralgic pains on the right side: subsequently she had complete paralysis of the fifth, and of the sixth, seventh, and eighth nerves of the right side. After death the right vertebral artery was found engaged in the mass of syphilitic deposit; it must be added, however, that the (superficial) origin of the fifth nerve was itself softened, opposite the pons. Another mode in which syphilitic disease very probably causes neuralgia of the fifth, in a certain number of cases, is by injuring the Gasserian ganglion, upon the integrity of which (according to Waller's general law concerning the ganglia of posterior nerve-roots) the nutrition of the sensory root of the trigeminus materially depends. I have seen an example (as I cannot but suppose) of this sequence of morbid events; the evidence appears sufficiently complete, although I was unable to obtain a *post mortem* examination:

W. M., a house painter, of extremely dissipated habits, but who had never suffered either from distinct symptoms of alcoholism, nor from any affection traceable to lead-poisoning. In March, 1867, he applied to me on account of neuralgic pain, affecting chiefly the right eyeball, but also darting along the course of the frontal nerve of that side; after a short time it extended also into the infra-orbital nerves. He bore several scars of tertiary ulcers about the nose and forehead, and made no secret of having suffered from chancre six or seven years before, and from subsequent secondary and tertiary symptoms. I was consequently not at all surprised at his developing severe iritis (right) after he had been a fortnight under my care, although I had from the first given large doses of iodide of potassium; but I was not prepared for the extensive processes of destruction which followed, notwithstanding that I immediately commenced mercurial treatment, and applied atropine. I remarked that while the inflammation of the iris proceeded with great violence, the cornea was also much more severely affected than is usually the case in syphilitic iritis; in fact, the changes closely resembled those which have been noted after section of the fifth at the Gasserian ganglion, and at the date of the patient's death (seventeen days from the commencement of the iritis) a corneal ulcer was on the point of perforating. For the first three or four days after the iritis set in, the neuralgic pains went on augmenting in intensity, and extended into all three divisions of the fifth; there was a copious discharge from the right nostril. Almost suddenly, on the fourth day, the pains abated and then ceased, and it was now evident that the whole surface of the right half of the face was completely anæsthetic. Two days later a dark-red patch appeared on the cheek, and in the course of the next two days this ulcerated, the ulcer presenting a somewhat livid appearance, and exuding a sanious discharge; at the same time, superficial ulcers appeared on the right side of the tongue, and coalesced to form one large sore. The sores both on cheek and tongue assumed more and more a gangrenous appearance, and on the sixteenth day from the commencement of iritis there was considerable loss of substance in both these situations. On the evening of this day (the patient having become extremely depressed and much emaciated) general epileptiform convulsions set in, and followed each other rapidly; in a few hours coma supervened, and the patient sank the next day. No *post mortem* could be obtained; but it seems extremely probable, from the above history, that the Gasserian ganglion was early involved in the syphilitic inflammation, and that the neuralgia and subsequent anæsthesia, the iritis, and the other trophic lesions, were due to the injury inflicted upon it.

The treatment of syphilitic pains will, in doubtful cases, often give us valuable asssurance of the correctness of our diagnosis. Where the disease is extensively diffused, we may fail to do any good; but, in cases where the syhilitic mischief is limited to a small portion of the meninges, we may often arrest it. In all merely suspicious cases, where the pain is thus limited, it will be well to use iodide of potassium tentatively—forty to sixty grains daily. But, where the pains are very severe and continuous, and there is danger to the integrity of the eye, or threatenings of a paralytic attack are observed, it is better not to trust to anything short of mercury, used in such a manner as just to stop short of absolute ptyalism. In very bad cases, like the last one narrated, we may fail to produce any good effect, but, where the specific treatment is commenced in good time, we may not unfrequently succeed in arresting the symptoms with a rapidity that assures us of the correctness of the diagnosis of syphilis.

CHAPTER VIII.

PAINS OF SUBACUTE AND CHRONIC RHEUMATISM.

So firmly is the idea of an essential connection between rheumatism and neuralgia implanted in the popular mind, and, indeed, in the minds of a certain portion of the medical profession, that the two complaints are continually confounded. In the great majority of instances, the mistake made is that of calling neuralgia a "rheumatism." But the opposite error occasionally occurs, and a patient is styled "neuralgic" who is really suffering from chronic rheumatism.

As true neuralgia is an essentially localized disease, there can be no excuse for mistaking for it the more typical cases of chronic rheumatism, in which a number of different joints, muscles, or tendons, are affected, more especially in the advanced stages, when the characteristic fixed contractions of the limbs and extremities have occurred. But there are a few cases in which, either with or without a previous history of acute rheumatism, one, or perhaps two, joints begin to suffer vague pains, which after a little time begin to shoot down the course of the limb, and are aggravated from time to time in a manner which superficially much resembles neuralgia; and when the malady has reached a certain intensity the pains may be so much more severely felt in the longitudinal axis of the limb than in the immediate neighborhood of a joint, that the patient forgets that in reality they commenced either within a joint (as the elbow or hip), or in the fibrous structures immediately outside it. Certain localties are much more frequently the seat of this kind of affection than other parts of the body; thus it occurs, perhaps in nine-tenths of the cases, in the neighborhood either of the shoulder (especially involving the insertions of the deltoid and triceps muscles), of the elbow (particularly affecting the tendinous insertions of the muscles on the internal aspect of the forearm), or the hip (extending to the aponeuroses on the outer and back part of the thigh): in all these cases there is a considerable superficial resemblance to true neuralgic pains. Nevertheless, the diagnosis need not present any serious difficulties after the earliest stages: for there soon arises a very diffuse and acute tenderness of the parts, and usually an amount of generalized swelling, which, though it may not be readily detectable by the eye, is sensible enough to the touch. Movement of the parts is also very painful; but usually not with the acute and agnozing pain which occurs in myalgia.

It is, however, upon signs which are of a more general character that we ought chiefly to rely for diagnosis. The fact that the patient has previously experienced a genuine attack of acute rheumatism, though of some value, is by no means to be taken as a conclusive argument that the present attack is of a rheumatic nature. The really important matter is, that whether the patient has or has not suffered acute rheumatism before the occurrence of the subacute or chronic form, the latter will always be attended by more or less of the specific constitutional disturbance of rheumatism. I would carefully abstain from the assumption that rheumatism is originally dependent on a blood-poisoning, a theory which I believe to be most doubtful and very probably false; but there is, nevertheless, a truly specific character about the general phenomena in acute rheumatism, and I maintain that similar though less-marked phenomena are always to be seen even in the mildest and least acute forms of rheumatism. Thus there will be, invariably, more or less of the peculiar sallow anæmia, together with red flushing of the cheeks when the pain is at the worst; and there will be a certain amount of the oily perspiration which makes the faces of rheumatic patients look shiny and greasy. No doubt these characteristics will sometimes be very slightly developed, but I believe that attentive observation will always discover them in any case which is genuinely rheumatic. One case, in particular, which has been under my care, very strongly impresses me with the value of these diagnostic signs, where otherwise the symptoms are obscure:

L. P., aged thirty-one, single, a printer by trade, applied to me, January, 1863, suffering from what I at first decidedly thought was cervico-brachial neuralgia, the pain having followed exposure to cold and wet, situated in the lower part of the neck, the shoulder, elbow and inner side of the right arm, and existing nowhere else. The character of the pain was described as at least remittent, if not distinctly intermittent. The pulse was not more than 78; the tongue was thickly coated with white fur, but the man did not complain of thirst, and there were no evident signs of fever. As the pains had only existed for about a fortnight, it appeared an excellent case for cure by the hypodermic injection of morphia; and, accordingly this was used in quarter-grain doses twice a day. After about ten days an attempt was made to do without the morphia, but the pains returned, worse than before, and meantime the tongue had remained uniformly coated, and was now very yellow; the appetite was bad, and there was some increase in frequency of pulse. It now struck me, for the first time, that the man presented, in a slight degree, the sallow and red tint and oily features of a rheumatic patient; it was now found that sweat and urine were distinctly acid. Acting on this idea, I administered five grains of iodide of potassium, and thirty grains of bicarbonate of potassium, four times every twenty-four hours, after giving a moderate saline aperient. The result was manifest improvement within twenty-four hours, and almost complete relief of the pain within three or four days (the urine never becoming distinctly alkaline, however.) As the attack subsided, the oily appearance of the skin disappeared, and the rheumatic tint was replaced by mere ordinary pallor, which the patient lost after taking a short course of steel.

At the time this case occurred to me, I was not aware of the importance, in doubtful instances, of looking to the temperature; but subsequent experience has convinced me that in every truly rheumatic case, however limited in extent, there is a real, though it may be a small, rise of temperature. The thermometer will be found to mark from $99\frac{1}{4}°$ to $100°$ Fahr., and this, joined with the appearances above mentioned, and a strong acidity of urine, will be sufficient to distinguish the complaint as rheumatic; and the striking effect of such remedies as iodide with bicarbonate of potash, followed up with sesquichloride of iron, in full doses, helps still further to distinguish the cases from true neuralgias. Since the introduction of the full doses of the iron-tincture in the treatment of acute rheumatism, I have had the opportunity of treating two of these cases of subacute rheumatism in the same manner, viz., with the iron from the first, and the results have been most satisfactory in every way. These cases were independent of a much larger number, treated in the same way, in which the symptoms of rheumatism were more generalized and more severe.

CHAPTER IX.

PAINS OF LATENT GOUT.

Pains which are connected with a chronic and more or less latent form of gout not unfrequently receive the designation "neuralgic," and are treated upon that erroneous theory of their pathology. I have already endeavored to show that there is by no means that intimate causal relation between gout and neuralgia which is very commonly assumed to exist: true neuralgia is, I believe, only caused in an indirect and secondary manner by the gouty condition setting up changes of the blood-vessels, which precipitate the occurrence of the neuralgic malady, to which the patient was otherwise predisposed from birth. But the common idea, both without and within the profession, seems to be that neuralgia is only one expression, and that a quite common one, of the gouty habit. Nevertheless, with strange inconsistence, the kind of truly gouty pains of which I am now speaking are constantly treated upon a special plan, upon the supposition that they are neuralgic.

There are six situations in which gouty pains are apt to be developed in a way to lead to the false diagnosis of neuralgia: (1) In the eye; (2) more indefinitely within the cranium; (3) in the stomach, simulating gastralgia; (4) in the chest, simulating angina pectoris; (5) in the dorsum of the foot, simulating neuralgia of the anterior tibial nerve; (6) in a somewhat diffuse manner about the hip and back of thigh, simulating sciatica.

It is not really a common thing to find such cases very difficult of diagnosis, provided that the possibility of their occurrence has been carefully noted; for the gouty habit has a number of slight manifestations which are usually enough to discover it even when its more decided symptoms are entirely wanting.

Thus, in the first place, it will be almost invariably found, on inquiry, that the patient has always been intolerant of beer and of sweet wines. Also, he has been liable (either after a single large excess in eating or a prolonged course of a diet too highly animalized in proportion to the amount of exercise taken) to attacks of general malaise, with or without uneasiness, just short of decided pain, about the metacarpo-phalangeal joint of the great-toe, and ending after a few hours or days with a free discharge of uric acid. Less frequently, but still very often, it will be found that he has some deposit of lithate of soda (chalk-stone) in some situation where its presence does not necessarily arrest attention; Dr. Garrod has shown how often these little tophi are found in the cartilage of the ear. Careful examination will sometimes detect their presence in the sclerotic of the eye. But in doubtful cases it would be always well to make a cautious trial of colchicum, which, if the case be gouty, will nearly always produce an amount of relief sufficient to confirm the the diagnosis of gout. At least, this rule holds goods for the external forms; but in the case of the supposed gouty pseudo-angina it is far best to trust to opium, as colchicum may prove too depressing to a heart which may quite possibly be already the subject of organic disease. My own impression is, that it was these cases of gouty heart-pain, which are not true angina at all, that procured for opium its high reputation for relieving the latter disease, a reputation which is by no means confirmed by my own experience, since I have found that drug enormously inferior to stimulants like ether in its power to relieve genuine angina.

Lastly, if there be no other possibility of making ourselves certain whether there is or is not a gouty taint at the bottom of the quasi-neuralgic pains, we may adopt Dr. Garrod's test of subjecting the serum of the blood to a search for uric acid (thread-test).

CHAPTER X.

COLIC, AND OTHER PAINS OF PERIPHERAL IRRITATION.

Colic, or painful half spasm, half paralysis of the large intestines, is the best example of a kind of spasmodic pains to which some authors accord the name of neuralgia, as it seems to me without good reason. They appear to be quite independent of the operation of the neurotic temperament, and to be caused entirely by the operation of some local irritant, or narcotic irritant, upon the muscular fibres of the viscus. In the case of colic this influence is most frequently and most powerfully exerted by lead, which undoubtedly becomes locally deposited in chronic poisoning with that metal; at other times it is produced by the irritation of indigestible food passing along the alimentary canal.

That there may be such a thing as enteralgia, of really neuralgic character, I do not deny; on the contrary, so far as regards the rectum, I have myself seen such a case. But true neuralgia of the large bowel is exceedingly uncommon; what goes by the name is usually either colic from local irritation of the viscus; or a mere hysterical hyperæsthesia of the lining membrane, which is one of the occasional phenomena of spinal irritation; or else it is a case of neuralgia of the abdominal wall, such as is included in the description of "lumbo-abdominal neuralgia," in Part I. of this work.

There is no occasion to describe minutely the symptoms of so familiar a disease as lead-colic, or as colic from irritation by indigestible food, when they occur in their typical forms. In the former case the marked constipation which ushers in the attack of pain, and the peculiar greenish-yellow sallowness nearly always seen in the countenance, ought to be sufficient to direct examination to the gums (for the blue line) and inquiry as to any possible impregnation of the system with lead, owing either to the nature of the patient's occupation, or to some accidental entry of the poison into the drinking-water, or its inhalation from the walls of newly-painted rooms, etc. In the latter case, the fact that the attack of colic was shortly preceded by a meal, either of obviously indigestible food, or too copious in quantity and heterogeneous in kind, or too hastily eaten without sufficient mastication, supplies a clew.

But there are a few cases representing minor degrees of either of these kinds of colic, that are much less easy to diagnose distinctly.

Lead-poison sometimes enters the system continuously, for a long period, but in proportions too minute to produce the effects which we identify as an attack of lead-colic. I believe that for the production of the latter complaint it is necessary that the poisoning shall be sufficiently intense completely to paralyze a considerable piece of bowel, thus altogether hindering peristalsis, or, rather, making the peristaltic acts of the non-paralyzed portions above worse than fruitless. But there is a minor degree in which it may happen that the local affection (owing, I believe, to a less extensive deposit of lead in the bowel) does not reach the decidedly paralytic stage; the state then is one of irregular and painful spasm of individual fibres (quite possibly intermingled with paralysis of a few others), and the practical result is irregularity of evacuation—now diarrhœa, and again constipation—and the frequent recurrence of twinges of pain that are easily mistaken for abdominal neuralgia. Such symptoms as these are nearly always found to have occurred, if proper inquiry be made, in those examples of chronic lead-poisoning in which the toxic process goes on to the development of epilepsy, or marked symmetrical paralysis of the wrist-extensors, without the patient having ever suffered an attack of ordinary colic. In these slow and insidious cases the constitutional affection may not have reached the height at which the complexion and general aspect of the patient suggests metallic poisoning: and the case may present very neuralgia-like features. The absence of the *points douloureux* is not, as we have seen, conclusive against neuralgia in its early stages. It is therefore an excellent rule, in all cases of chronic recurrent spasmodic pain in the abdomen, especially in men, to investigate the possibilities of lead-poisoning; and, if the slightest suspicious appearance of the gums be found, this track of inquiry must be followed up exhaustively before we abandon the idea. The absence of all special neurotic history in a patient's family should increase our suspicions respecting pains of this character that continue with an obstinacy which makes it unlikely they are due to improper food.

Pains of abdominal irritation are, however, without doubt produced in some cases by unsuspected faults of diet, and may even recur in such a quasi-periodic manner as to strongly suggest the idea of neuralgia in the lumbo-abdominal nerve. One special variety of this happens, I believe, much more often than is thought. A patient will habitually take considerable quantities of some article of food which he does not readily digest, but which is not at all acutely irritant: under these circumstances a simple accumulation is apt to take place in the colon, especially at the top of the ascending colon, the top of the descending colon, or just above the sigmoid flexure, or else in the cæcum. The result of accumulation in the last of these places is not unfrequently typhlitis and perityphlitis, this part of the bowel having (for some reason) a special tendency to inflammation. Deposits in the other localities named are rarely the cause of inflammation, but they very frequently give rise to violent pain, which is exceedingly apt to be taken for the pain either of gall-stone, of renal calculus, or else of some abdominal neuralgia. In cases, therefore, where there is any possibility that accumulation is the cause of pain, it is highly desirable to commence with a dose of castor-oil and laudanum, followed up, if needful, by the administration of a large warm-water enema, given through an O'Beirne's tube. The most violent and recurrent attacks of pain in the renal region, the flank, the abdomen, or the groin, will sometimes be instantly cured by such means, sufficiently proving the non-neuralgic character of the complaint.

I have elsewhere explained that the impaction of a renal or an hepatic calculus, in the ureter or the ductus choledochus, may set up a true neuralgia in persons with the requisite congenital predisposition. The passage of renal or hepatic calculi may give rise to symptoms falsely suggesting neuralgia, which require just to be mentioned here. But there is no need to dwell much upon the diagnosis, for the passage of renal or hepatic calculi has always attendant symptoms and features of constitutional history, which ought to preserve the physician from mistake. The sensation of constriction, of nausea and vomiting, the faintness approaching to collapse, the persistent and constantly increasing severity of the pain up to the moment at which mechanical relief occurs, to say nothing of other phenomena, are distinctive to the skilled observer, and, when taken in conjunction with the history of past attacks, if any, will always prevent mistakes. In the few cases which might still be doubtful it will be well to try the the effect of a relaxing dose of chloroform, which, in the case of calculus, will often put an end to the paroxysm at once and finally.

CHAPTER XI.

DYSPEPTIC HEADACHE.

A final word or two must be given to the distinction between neuralgia of the head and an affection so utterly different that it is surprising that they should be so frequently confounded. One constantly hears medical men speak of "sick headache" (migraine) as if it were the same thing as headache from indigestion; and, unfortunately, they often treat migraine upon this confused and erroneous notion, doing no little mischief thereby.

But, although migraine. already amply described, is entirely independent of the state of digestion, and its stomach-phenomena are purely secondary to the affection of the fifth nerve, there is a kind of headache really dependent on imperfect digestion. The sufferers from these headaches are dyspeptics whose stomach troubles are the result of chronic gastric catarrhal inflammation. (In the acute form of gastric catarrh there are even more severe headaches; but the general symptoms of the disorder are too marked to allow us to mistake the case for neuralgia complicated with secondary stomach disturbance.) The patients in question have frequently passed so gradually into the dyspeptic condition as to have become accustomed to it, and inclined to forget that the stomach was the organ which first gave them annoyance. The headaches, which occur from time to time, are either frontal or (more frequently) occipital in position, and they are usually quite evenly bilateral; still, there is not enough uniformity of difference between them and true migraine, in this respect, to enable us to establish a decided diagnosis upon it. This much may be said, however: that the pain is rarely or never seated in one parietal region, as is frequently the case with migraine and with clavus. The patient suffers very strikingly, in almost every case, from languor and a feeling of inability to exert himself; and has also much aching pain in the limbs, and usually a pain (sometimes very severe) in the scapular region. The tongue may vary a good deal in appearance, especially as regards the degree of general redness; but it always has enlarged papillæ, most prominent toward the tip, and more or less thick furring at the back, and reaching forward, in some cases, nearly to the tip, to which the "strawberry" aspect is then confined. The headache is frequently joined with nausea, but never with absolute vomiting, unless the stomach has been provoked with a meal that gives it more trouble than usual. The desponding frame of mind which this kind of dyspeptics always exhibit distinguishes them, in most cases, quite sufficiently (together with the unwholesome complexion, the appearance of the tongue, and the great complaints of general malaise and aching and feebleness of the limbs) from the victims of migraine, who are often persons of bright spirits and lively intelligence in the intervals of their attacks; but, above all, there is nothing of the regular and characteristic sequence of events which distinguishes the attacks of migraine. The attacks are not periodic, but nearly always depend on some chance dietary indiscretion, or other imprudence, which has visibly aggravated the stomach irritation. And, when the pain does come on, it has no uniform tendency to go on intensifying for some hours and culminate in vomiting, followed by sleep, after which the patient is free. On the contrary, the digestive disturbance is the provocation, and the pain itself is of a heavy character, with a sense of tension or fulness, and it does not go on intensifying in a regular manner, up to a climax, but hangs about in a dull, tormenting way, and frequently is just as bad after sleep as it was before. The diagnosis of these headaches from neuralgic headache is not really difficult; it only requires the use of a fair amount of caution in observation. It would, however, be exceedingly advantageous that the word "sick-headache" should be dropped altogether, and that migraine should always be called by that name (or "megrim," if you will), and that headaches really proceeding from chronic catarrhal disease of the stomach should be called "dyspeptic" headaches. The present state of nomenclature does much to perpetuate a confusion of ideas which ought not to exist any longer, and which leads to much practical mischief.